Aerosol Particulate Separation Theory and Technology

气溶胶粒子分离理论与技术

向晓东　常玉锋　编著

Beijing
Metallurgical Industry Press
2020

内 容 提 要

本书共分为9章，第1、2章分别论述了气溶胶粒子的基本性质和在不同外力作用下气溶胶粒子的运动行为。第3、4章分别论述了对气溶胶粒子分离技术进步具有重要指导作用的扩散与凝并理论。第5~8章分别论述了气溶胶粒子的空气动力分离、静电捕集、纤维过滤和湿式净化的原理及应用。第9章介绍了目前最具代表性和应用价值的气溶胶粒子复合高效净化技术，阐述了各复合净化装置的结构特征、技术核心、分离机理和应用优势。

本书既可作为环境、安全、土木、化工、冶金、建筑等相关专业的硕士或博士生双语教材和科研工具书，也可作为大气污染控制领域的科研和工程技术人员的专业参考书。

图书在版编目(CIP)数据

气溶胶粒子分离理论与技术/向晓东，常玉锋编著.
—北京：冶金工业出版社，2020.8
ISBN 978-7-5024-8521-4

Ⅰ.①气… Ⅱ.①向… ②常… Ⅲ.①气溶胶—研究
Ⅳ.①O648.18

中国版本图书馆CIP数据核字(2020)第138750号

出 版 人 陈玉千
地　　址 北京市东城区嵩祝院北巷39号 邮编 100009 电话 (010)64027926
网　　址 www.cnmip.com.cn 电子信箱 yjcbs@cnmip.com.cn
责任编辑 俞跃春 刘林烨 美术编辑 郑小利 版式设计 孙跃红
责任校对 郑 娟 责任印制 禹 蕊
ISBN 978-7-5024-8521-4
冶金工业出版社出版发行；各地新华书店经销；北京兰星球彩色印刷有限公司印刷
2020年8月第1版，2020年8月第1次印刷
787mm×1092mm 1/16；13印张；312千字；189页
69.00元

冶金工业出版社 投稿电话 (010)64027932 投稿信箱 tougao@cnmip.com.cn
冶金工业出版社营销中心 电话 (010)64044283 传真 (010)64027893
冶金工业出版社天猫旗舰店 yjgycbs.tmall.com
(本书如有印装质量问题，本社营销中心负责退换)

Preface

Haze, a phenomenon of polluted weather, is caused mainly by the aerosol particulate which is generated from burning fossil fuels such as coal and oil fuels. Today, the haze pollution due to the emission of fine particles is still severe in China. The fine particle pollutant source control is one of the best way to solve the haze problem by means of aerosol science and technology. However, because of the influence factors of the ambient medium, geometric shape, size distribution, physical and chemical properties of the particulates which are produced by different industrial processes being complex and various, it is usually quite difficult to meet the needs of the particulate pollutant emission standard if the traditional particle removing technology is applied.

To propel the particulate pollution control technology, the books of Xiang, Particle Collection Theory and Technology (Metallurgical Industry Press, 2002), and Fabric Filtration Theory, Technology, and Application (Metallurgical Industry Press, 2007) have been taken an active promotion effect on the field of particulate pollution control. Particle Collection Theory and Technology has been used as a text book for the postgraduate students of the environment and safety majors of our university and some other universities. But in the practice of education, researches, and applications, it is found that the aerosol science is the only internal motivation of the particulate pollutant control technology. As for the supplementary reading material of the particulate pollution control knowledge system, the book of Fundamentals of Aerosol Science and Technology (China Environmental Science Press, 2012) was dedicated by Xiang.

As the progress of aerosol particle separation technology, the implementation of national blue-sky plan, and the expansion of international communication and cooperation, our goal here is going to culture the well educated experts who are capable of application, innovation, and academic exchange as soon as possible in the field of the particulate pollution control.

Therefore, Aerosol Particulate Separation Theory and Technology. which is integrated aerosol science with traditional particle collection technology and some typical achievements of particulate pollution control researches in recent years, can be used for an extracurricular reading material at graduate level, for master or doctor degree candidates in safety, environmental, chemical, civil, mechanical, and architecture engineering as well as well as for the scientific researchers and technicians in the fields related to air pollution control.

Here, we would like to gratefully acknowledge Graduate School of Wuhan University of Science and Technology, State Environmental Protection Key Laboratory of Mineral Metallurgical Resources Utilization and Pollution Control, and National Key Research and Development Project (No. 2017YFC0210404) for their financial supports.

X. D. Xiang, Y. F. Chang

December 25, 2019

前　言

化石燃料燃烧产生的气溶胶微粒是导致大气污染的主要原因之一。目前，微细颗粒物排放所造成的雾霾污染状况依然严峻，控制雾霾的一条重要途径是运用气溶胶科学与技术进行微细颗粒物排放的源头控制。然而，不同工业过程产生的细颗粒物的传输介质、几何形态、粒径分布、物理化学性质等影响因素复杂多变，采取传统的除尘技术实现颗粒污染物达标排放通常是有难度的。

为推动颗粒物控制技术的发展，作者先后编著了《现代除尘理论与技术》（冶金工业出版社，2002 年出版）、《纤维过滤理论、技术及应用》（冶金工业出版社，2007 年出版），在国内颗粒污染物控制领域产生了积极影响。其中，《现代除尘理论与技术》自出版以来一直作为我校和其他部分院校安全和环境类专业的研究生教学用书，在人才培养、学科建设方面起到了促进作用。但在教学、科研与工程实践中逐渐发现，气溶胶科学才是除尘技术创新与发展的原动力。于是，作者编著了《气溶胶科学技术基础》（中国环境科学出版社，2012 年出版），作为颗粒污染物控制理论体系完备性的一个补充。

随着气溶胶微粒净化技术的进步、国家蓝天计划的实施、大气污染控制领域国际交流与合作的不断深化，培养适应气溶胶微粒净化技术发展的高素质创新人才显得日益重要。为此，将气溶胶科学与除尘技术整合，并引入近几年国内外在该领域有代表性的最新研究成果，编著一本知识体系完整、理论与实践融合的研究生英语教材很有必要。

本书适合于高等院校全英语和双语教学，对提高学生的科研能力、创新能力、英语科技写作与交流能力具有突出的教育意义。

本书的出版与研究得到了武汉科技大学研究生院、国家环境保护矿冶资源利用与污染控制重点实验室、国家重点研发计划项目（No. 2017YFC0210404）的支持与资助，在此表示感谢！

向晓东、常玉锋

2019 年 12 月 25 日

Contents

1 **Basic Nature of Particulate Pollutants** ······ 1
1.1 Particle Size ······ 1
1.1.1 Particulate Pollutants ······ 1
1.1.2 Equivalent Particle Diameters ······ 3
1.2 Size Distribution ······ 5
1.2.1 Particle Number Fraction ······ 5
1.2.2 Particle Frequency Distribution ······ 7
1.2.3 Particle Cumulative Frequency Distribution ······ 8
1.2.4 Particle Number and Mass Distribution ······ 10
1.3 Particle Concentration ······ 11
1.3.1 Relation of Number and Mass Concentrations ······ 11
1.3.2 Relation of Standards and Particulate Pollutant Concentration ······ 13
1.3.3 Relation of Human Health and Particulate Pollutant ······ 14
Exercises ······ 16
References ······ 17

2 **Motion of Aerosol Particulate in an External Force Field** ······ 18
2.1 Drag Force on a Single Spherical Particle ······ 18
2.1.1 Laminar Regime ······ 19
2.1.2 Transition Regime ······ 19
2.1.3 Turbulent Regime ······ 19
2.1.4 Particles too Small for Stokes' Law ······ 20
2.2 Motion of a Particle in an External Force Field ······ 20
2.2.1 Motion of a Particle in Gas under Gravity ······ 20
2.2.2 Motion of a Particle under a Centrifugal Force ······ 23
2.2.3 Motion of a Charged Particle in an Electric Field ······ 24
2.3 Suspension of a Particle in a Duct ······ 25
2.3.1 Particle Suspension Caused by a Shear Lift Force ······ 26
2.3.2 Particle Suspension Caused by Pressure Difference ······ 26
Exercises ······ 30
References ······ 31

3 **Diffusion of Aerosol Particulate** ······ 32
3.1 Basic Diffusion Law ······ 32

3. 1. 1 Fick' s First Law ···· 32
3. 1. 2 Fick' s Second Law ···· 33
3. 2 Diffusion in Still Gas ···· 33
3. 2. 1 Diffusion for Reflection Wall in Still Gas ···· 33
3. 2. 2 Diffusion for Absorption Wall in Still Gas ···· 36
3. 2. 3 Diffusion for Absorption Surface of a Sphere in Still Gas ···· 38
3. 2. 4 Diffusion for Absorption Surface of a Cylindrical Tube in Still Gas ···· 40
3. 3 Diffusion in Static Gas Flow ···· 42
3. 3. 1 Diffusion for Reflection Wall of a Rectangular Duct ···· 42
3. 3. 2 Diffusion for Reflection Wall of a Cylindrical Duct ···· 44
3. 3. 3 Diffusion for Absorption Wall of a Rectangular Duct ···· 44
3. 3. 4 Diffusion for Absorption Wall of a Cylindrical Duct ···· 45
3. 4 Diffusion of a Gas Flow Around an Axisymmetric Body ···· 46
3. 4. 1 Diffusion of a Gas Flow Around a Cylinder ···· 46
3. 4. 2 Diffusion of a Gas Flow Around a Sphere ···· 49
Exercises ···· 50
References ···· 51

4 Coagulation of Aerosol Particulate ···· 52
4. 1 Brownian Coagulation ···· 52
4. 1. 1 Brownian Coagulation of Monodisperse Particles ···· 52
4. 1. 2 Brownian Coagulation of Polydisperse Particles ···· 54
4. 2 Electrical Coagulation ···· 57
4. 2. 1 Coulomb' s Coagulation ···· 57
4. 2. 2 Electrostatic Coagulation in an Alternating Electric Field ···· 59
4. 3 Particle Distribution in Coagulation Process ···· 60
4. 3. 1 Assumption of Self-preserving ···· 60
4. 3. 2 Particle Size Distribution Simplification in Coagulation Process ···· 62
Exercises ···· 64
References ···· 65

5 Aerodynamic Separation of Particulate ···· 67
5. 1 Settling Chamber ···· 67
5. 1. 1 Settling Chamber of Laminar Flow ···· 67
5. 1. 2 Settling Chamber of Turbulent Flow ···· 69
5. 2 Inertial Separators ···· 71
5. 2. 1 Inertial Deposition in Arch Duct ···· 71
5. 2. 2 Cascade Impactor ···· 74
5. 3 Cyclone Collector ···· 77

5. 3. 1 Flow Field of Cyclone ········· 77
5. 3. 2 Collection Efficiency of Cyclone ········· 79
5. 3. 3 Pressure Drop of Cyclone ········· 82
5. 3. 4 Dimensions of Cyclone ········· 83
5. 3. 5 Multiple Cyclone ········· 84
Exercises ········· 87
References ········· 89

6 Electrostatic Precipitation ········· 91
6. 1 Basic Principles of ESP ········· 91
6. 1. 1 Types of ESP ········· 91
6. 1. 2 Particulate Collection Process of ESP ········· 92
6. 2 Particle Charging ········· 93
6. 2. 1 Corona and Ion Generation in ESP ········· 93
6. 2. 2 Charge on a Particle ········· 94
6. 3 Electric Field ········· 95
6. 3. 1 Electric Field in Wire-tube ESP ········· 95
6. 3. 2 Electric Field in Wire-plate ESP ········· 98
6. 4 Deutsch Equation ········· 102
6. 4. 1 Collection Efficiency of Wire-plate ESP ········· 102
6. 4. 2 Collection Efficiency of Wire-tube ESP ········· 103
6. 5 Effect Factors on Collection Performance of ESP ········· 104
6. 5. 1 Charging Electric Field Strength ········· 104
6. 5. 2 Collection Electric Field Strength ········· 105
6. 5. 3 Dust Re-entrainment ········· 106
6. 5. 4 Specific Collection Area ········· 108
6. 5. 5 Voltage-current Characteristics ········· 108
6. 5. 6 Dust Resistivity ········· 111
6. 5. 7 Gas Distribution ········· 113
6. 6 ESP Design ········· 113
6. 6. 1 Information Required for an ESP Design ········· 113
6. 6. 2 Overall Mass Collection Efficiency ········· 115
6. 6. 3 Design Procedure ········· 116
Exercises ········· 118
References ········· 119

7 Fabric Filtration ········· 121
7. 1 Mechanisms of Particle Capture by a Fiber ········· 121
7. 1. 1 Interception ········· 122

7.1.2 Inertial Impaction ········ 123
7.1.3 Diffusion ········ 124
7.1.4 Combination Collection Efficiency of Independent Mechanisms ········ 124
7.2 Collection Efficiency of a Fibrous Filter Bed ········ 125
7.2.1 Depth Filtration Efficiency of a Filter ········ 125
7.2.2 Surface Filtration Efficiency of a Filter ········ 127
7.3 Pressure Drop of a Fibrous Filter Bed ········ 127
7.3.1 Pressure Drop of a Clean Filter Bed ········ 127
7.3.2 Pressure Drop of a Filter Bed with Cake Filtration ········ 129
7.4 Filter Media ········ 130
7.4.1 General Description ········ 130
7.4.2 Membrane Filter ········ 130
7.5 Industrial Bag-house ········ 134
7.5.1 Pulse Jet ········ 134
7.5.2 Mechanical Shaking ········ 135
7.5.3 Backwash ········ 135
7.5.4 Sonic ········ 136
7.6 Bag-house Design ········ 137
7.6.1 Filter Emissions ········ 137
7.6.2 Filtration Velocity ········ 138
7.6.3 Cleaning Design ········ 139
Exercises ········ 144
References ········ 145

8 Scrubbers ········ 147
8.1 Mechanisms of Particle Capture of a Single Droplet ········ 147
8.1.1 Interception ········ 147
8.1.2 Inertial Impaction ········ 149
8.1.3 Diffusion ········ 149
8.1.4 Combination Collection Efficiency of Independent Mechanisms ········ 150
8.2 Overall Efficiency by Monodisperse Drops ········ 150
8.2.1 Cylindrical Model ········ 150
8.2.2 Box Model ········ 151
8.3 Overall Efficiency by Drops with Size Distribution ········ 153
8.4 Performance of Scrubbers ········ 155
8.4.1 Spray Tower ········ 155
8.4.2 Packed Tower ········ 158
8.4.3 Venturi Scrubber ········ 161
Exercises ········ 164

References ········· 164

9 Hybrid Separators ········· 166
9.1 Aerodynamic Electrostatic Precipitators ········· 166
9.1.1 Transverse Plate ESP ········· 166
9.1.2 Electrostatic Enhancement Cyclone ········· 170
9.2 Electrostatic Enhancement Fabric Filters ········· 171
9.2.1 Unipolar Dust Pre-chargers for Fabric Filters ········· 172
9.2.2 Bipolar Dust Pre-chargers for Fabric Filters ········· 175
9.3 Electrostatic Scrubbers ········· 176
9.3.1 Collection Efficiency Prediction ········· 176
9.3.2 Electrically Charged Spray Tower ········· 178
9.3.3 Wet Electrostatic Precipitators ········· 178
9.4 Aerodynamic Free Rotary Thread Scrubber ········· 180
9.4.1 Working Principal of Free Rotary Thread Scrubber ········· 180
9.4.2 Collection Efficiency Prediction ········· 181
Exercises ········· 187
References ········· 187

目　录

1　颗粒污染物的基本性质 …… 1
1.1　粒子大小 …… 1
1.1.1　颗粒污染物 …… 1
1.1.2　等效粒径 …… 3
1.2　粒径分布 …… 5
1.2.1　粒子数量分数 …… 5
1.2.2　粒子频率分布 …… 7
1.2.3　粒子累积频率分布 …… 8
1.2.4　粒子数量与质量分布 …… 10
1.3　粒子浓度 …… 11
1.3.1　粒子数量浓度与质量浓度的关系 …… 11
1.3.2　标准与颗粒污染物浓度的关系 …… 13
1.3.3　人类健康与颗粒污染物的关系 …… 14
习题 …… 16
参考文献 …… 17

2　外力作用下气溶胶粒子的运动 …… 18
2.1　单个球形粒子在气体中的阻力 …… 18
2.1.1　层流区 …… 19
2.1.2　过渡区 …… 19
2.1.3　湍流区 …… 19
2.1.4　微细粒子的斯托克斯阻力修正 …… 20
2.2　外力场中单个粒子的运动 …… 20
2.2.1　重力场中粒子的运动 …… 20
2.2.2　离心力场中粒子的运动 …… 23
2.2.3　电流场中粒子的运动 …… 24
2.3　管道流中粒子的悬浮 …… 25
2.3.1　剪切力引起的粒子悬浮 …… 26
2.3.2　压力差引起的粒子悬浮 …… 26
习题 …… 30
参考文献 …… 31

3　气溶胶粒子的扩散 …… 32
3.1　扩散基本定律 …… 32

3.1.1　费克第一定律……32
3.1.2　费克第二定律……33
3.2　静止气体中气溶胶粒子的扩散……33
3.2.1　静止气体中气溶胶对反射壁的扩散……33
3.2.2　静止气体中气溶胶对吸收壁的扩散……36
3.2.3　静止气体中气溶胶对球形吸收表面的扩散……38
3.2.4　静止气体中气溶胶对圆柱吸收表面的扩散……40
3.3　定常流中气溶胶粒子的扩散……42
3.3.1　气溶胶对矩形管道反射壁表面的扩散……42
3.3.2　气溶胶对圆形管道反射壁表面的扩散……44
3.3.3　气溶胶对矩形管道吸收壁表面的扩散……44
3.3.4　气溶胶对圆形管道吸收壁表面的扩散……45
3.4　定常流中气溶胶对封闭体的绕流扩散……46
3.4.1　绕圆柱体流动的气溶胶粒子扩散……46
3.4.2　绕球体流动的气溶胶粒子扩散……49
习题……50
参考文献……51

4　气溶胶粒子的凝并……52
4.1　布朗凝并……52
4.1.1　单分散性粒子的布朗凝并……52
4.1.2　多分散性粒子的布朗凝并……54
4.2　电凝并……57
4.2.1　库伦凝并……57
4.2.2　交变电场中的静电凝并……59
4.3　凝并过程中气溶胶粒子的粒径分布……60
4.3.1　自保假设……60
4.3.2　凝并过程中粒径分布简化……62
习题……64
参考文献……65

5　气溶胶粒子的空气动力分离……67
5.1　重力沉降室……67
5.1.1　层流重力沉降室……67
5.1.2　湍流重力沉降室……69
5.2　惯性分离器……71
5.2.1　弧形管道中的惯性沉降……71
5.2.2　垂直板惯性冲击分离器……74
5.3　旋风除尘器……77

5.3.1　旋风除尘器内流畅 …… 77
5.3.2　旋风除尘器收集效率 …… 79
5.3.3　旋风除尘器压力损失 …… 82
5.3.4　旋风除尘器几何尺寸比 …… 83
5.3.5　多管旋风除尘器 …… 84
习题 …… 87
参考文献 …… 89

6　静电除尘器 …… 91
6.1　静电除尘原理 …… 91
6.1.1　静电除尘器的类型 …… 91
6.1.2　静电除尘过程 …… 92
6.2　粒子荷电 …… 93
6.2.1　电晕现象与离子发生 …… 93
6.2.2　粒子的荷电量计算 …… 94
6.3　电场 …… 95
6.3.1　线管式静电除尘器的电场 …… 95
6.3.2　线板式静电除尘器的电场 …… 98
6.4　多依奇公式 …… 102
6.4.1　线板式静电除尘器的收集效率 …… 102
6.4.2　线管式静电除尘器的收集效率 …… 103
6.5　静电除尘器收集性能的影响因素 …… 104
6.5.1　荷电场强 …… 104
6.5.2　收尘场强 …… 105
6.5.3　二次扬尘 …… 106
6.5.4　比收尘面积 …… 108
6.5.5　伏安特性 …… 108
6.5.6　粉尘比电阻 …… 111
6.5.7　气流分布 …… 113
6.6　静电除尘器的设计 …… 113
6.6.1　静电除尘器设计参数 …… 113
6.6.2　总质量收集效率 …… 115
6.6.3　设计步骤 …… 116
习题 …… 118
参考文献 …… 119

7　纤维过滤 …… 121
7.1　孤立纤维的粒子收集机理 …… 121
7.1.1　拦截 …… 122

7.1.2　惯性碰撞 …… 123
7.1.3　扩散 …… 124
7.1.4　各单独机理的复合收集效率 …… 124
7.2　纤维过滤床层的过滤效率 …… 125
7.2.1　过滤器的内部过滤效率 …… 125
7.2.2　过滤器的表面过滤效率 …… 127
7.3　纤维过滤床层压力损失 …… 127
7.3.1　洁净过滤床层压力损失 …… 127
7.3.2　粉饼沉积的过滤床层压力损失 …… 129
7.4　过滤器的过滤介质 …… 130
7.4.1　概述 …… 130
7.4.2　覆膜滤料 …… 130
7.5　工业袋式除尘器 …… 134
7.5.1　脉冲袋式除尘器 …… 134
7.5.2　机械振打袋式除尘器 …… 135
7.5.3　反吹风袋式除尘器 …… 135
7.5.4　声波清灰袋式除尘器 …… 136
7.6　袋式除尘器设计 …… 137
7.6.1　排放浓度 …… 137
7.6.2　滤料过滤风速 …… 138
7.6.3　清灰设计 …… 139
习题 …… 144
参考文献 …… 145

8　洗涤器 …… 147
8.1　孤立液滴的粒子捕集机理 …… 147
8.1.1　拦截 …… 147
8.1.2　惯性碰撞 …… 149
8.1.3　扩散 …… 149
8.1.4　各单独机理的复合收集效率 …… 150
8.2　单分散性液滴的粒子捕集总效率 …… 150
8.2.1　圆柱模型 …… 150
8.2.2　箱式模型 …… 151
8.3　具有粒径分布液滴的粒子捕集总效率 …… 153
8.4　洗涤器性能 …… 155
8.4.1　喷淋塔 …… 155
8.4.2　填料塔 …… 158
8.4.3　文丘里洗涤器 …… 161
习题 …… 164

参考文献…………………………………………………………………………………… 164

9　复合机理分离器 ……………………………………………………………………… 166
9.1　空气动力静电除尘器 ……………………………………………………………… 166
9.1.1　横向极板静电除尘器 …………………………………………………………… 166
9.1.2　静电增强旋风除尘器 …………………………………………………………… 170
9.2　静电增强纤维过滤器 ……………………………………………………………… 171
9.2.1　单极预荷电静电增强过滤器 …………………………………………………… 172
9.2.2　双极预荷电静电增强过滤器 …………………………………………………… 175
9.3　静电洗涤器 ………………………………………………………………………… 176
9.3.1　粒子收集效率估算 ……………………………………………………………… 176
9.3.2　荷电水雾喷淋塔 ………………………………………………………………… 178
9.3.3　湿式静电除尘器 ………………………………………………………………… 178
9.4　空气动力自由旋线洗涤器 ………………………………………………………… 180
9.4.1　自由旋线洗涤器的工作原理 …………………………………………………… 180
9.4.2　自由旋线洗涤器的粒子收集效率估算 ………………………………………… 181
习题…………………………………………………………………………………………… 187
参考文献……………………………………………………………………………………… 187

1 Basic Nature of Particulate Pollutants

Airborne particulate, also called aerosol, is defined as the suspension particles of liquid or solid in air. Airborne particulate matter is a complex mixture of many different chemical species, originating from a variety of sources. The main source of airborne particulate matter is the combustion of fossil fuels, such as coal burning in power plant. Particulate matter is also emitted considerably from the processing and manufacturing industry, as well as from the motor vehicles.

Particles in the atmosphere consist of solid particles, liquid droplets, and liquid components contained within the solid particles. Particles are variable in relation to their concentration and their physicochemical and morphological characteristics. The particles in the air over a city cause the haze.

Particles in the atmosphere are either primary or secondary in nature. Primary particles are emitted or injected directly into the atmosphere from the pollutant sources, whereas secondary particles formed by aggregation or nucleation from gas-phase molecules. These secondary particles formed mostly from hydrocarbons, oxides of nitrogen, and oxides of sulfur in the atmosphere by gas-phase photochemistry mechanism. In this text book we just discuss the emission control of the primary particles.

1.1 Particle Size

1.1.1 Particulate Pollutants

Particulate pollutants are not chemically uniform, but rather come in a wide variety of sizes, shapes, and chemical compositions. All aerosol properties depend on particle size. Some are much more harmful to health, property, and visibility than others[1,2]. If particles are all the same in size, an aerosol is termed 'monodisperse'. Monodisperse particulate is extremely rare in nature. Generally, particles vary in size, called 'polydisperse'.

Aerosols range in size from 0.001μm to 100μm, so the particle sizes span several orders of magnitude, ranging from almost macroscopic down to near molecular sizes. Fig. 1.1 presents an overview of size range of the different particulates[3].

There are various types of aerosol, which are classified according to physical forms and ways of generation. The commonly used terms are 'dust', 'fume', 'smoke', 'fly ash', 'fog' and 'mist'.

Dusts are solid particles formed by crushing, grinding, sieving, powder loading, discharge or other mechanical action resulting in physical disintegration of a parent material. These particles

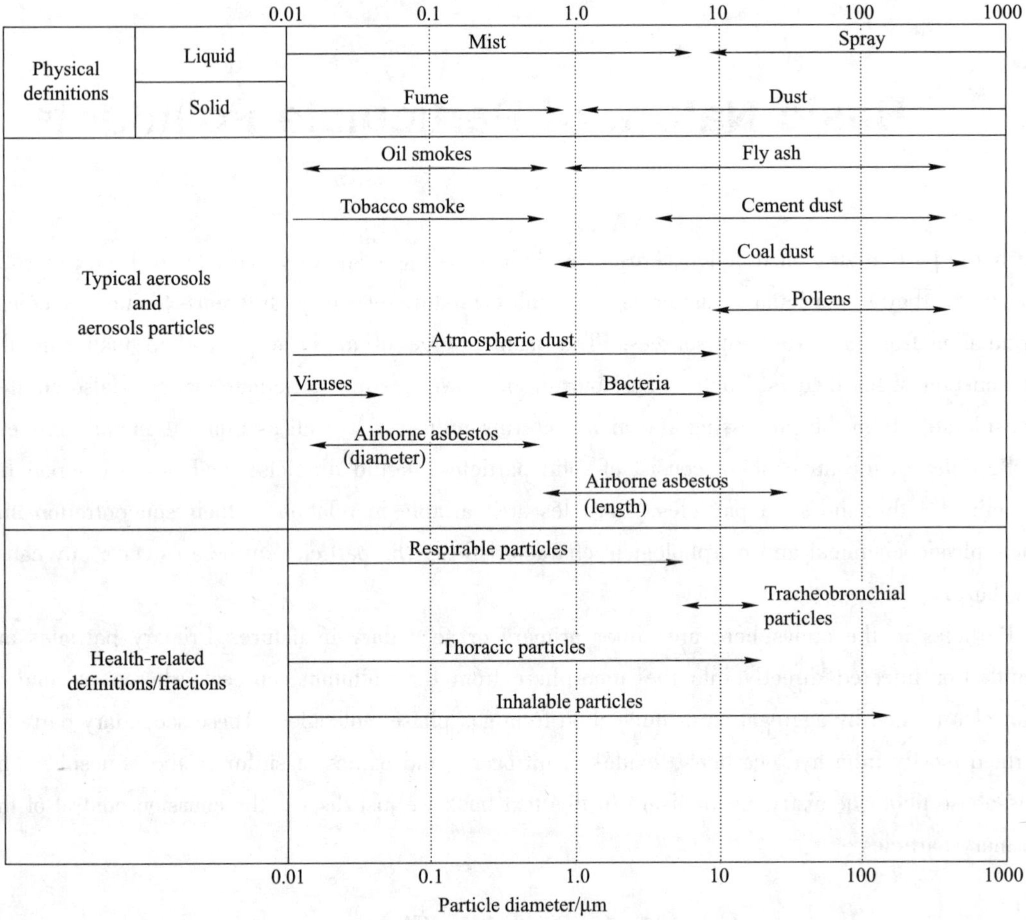

Fig. 1. 1　Particle size range for aerosols
(From Ian Colbeck and Mihalis Lazaridis, 2014)

have irregular shapes and have a size range from 1μm to 100μm.

Smoke and fume have a little difference in mining. Both are solid or liquid aerosol, created by combustion, evaporation, or condensation processes. The formation of tobacco smoke is one example which the reader has most likely personaly observed. The smoke consists of droplets of condensed hydrocarbons (oils, tars). When the temperature is high enough, these hydrocarbons are transparent gaseous molecules. As the combustion gases rise, they mix with colder air and reach the condensation temperature, at which the hydrocarbon gases forming the very small drops that make the visible smoke.

The puff of smoke is called fume. If the smoke is discharged in to the atmosphere relative stably and orderly from an exhaust pipe of an oil burning car or from a stack of coal burning boiler, we sometime say this is fume. The smoke produced by the flammable substance combustion in open area is not called fume, such as the smoke from a campfire or from a burning cigarette. The size range of smoke and fume is 0. 01μm to 1μm.

Most fuels contain some incombustible materials, which remain behind when the fuel is burned, called ash or fly ash. The ash left behind by combustion of wood, coal, or charcoal contains mostly the oxides of silicon, calcium and aluminum, with traces of other minerals. If the fuel is finely ground (or produced as a spray of fine droplets) and then burned, the remaining unburned ash particles may be quite small. The ash particles in the exhaust gases consist of two groups. One group has an average diameter of about 0.02μm. The other about 10μm. the smaller particles contain a much higher percentage of the more volatile materials in the ash than the larger particles. Almost certainly the fine particles are formed by condensation, in the furnace, of materials that have been vaporized during the combustion process. The larger particles are formed from the remaining mineral in the fuel which is not vaporized.

Fog or mist is the liquid droplets in air. These droplets are usually formed by condensation of supersaturated vapor of water or by spray of water physically. The size ranges of mist are wide. For example, in a wet method air pollution control system, it can be seen that a white plume is discharged in to the atmosphere from a stack. This white plume contains not only water droplets but also fine particulate[4]. Near the outlet of the stack, the fog size is larger than about 10μm. When the white plume goes up in air, the droplets are getting smaller to become haze due to the dispersion and evaporation. The size range of haze is 0.01μm to 10μm.

1.1.2 Equivalent Particle Diameters

Particle size is the most important descriptor to predict the behavior of the particles and the performance of the particulate pollutant control equipment. When particles are spherical, their radius or diameter can be used to describe their size. However, most particles are not spherical except when particles are liquid droplets. Therefore, equivalent diameter is commonly used. There are four equivalent diameters for irregular particle which are mainly used, including Martin diameter, volume equivalent diameter, Stokes diameter, and aerodynamic diameter.

1.1.2.1 Length Equivalent Diameter

Length equivalent diameter, sometimes called as Martin diameter[5], is the length of a particle in a particular direction under a microscope. To uniquely define length parallel to the reference direction and to provide an easily workable process at the same time, the length in question will be measured through a point midway between the top-most and the bottom-most points of the particle as it is oriented relative to the reference direction. This equivalent diameter is equal to distance AB, illustrated schematically in Fig. 1.2.

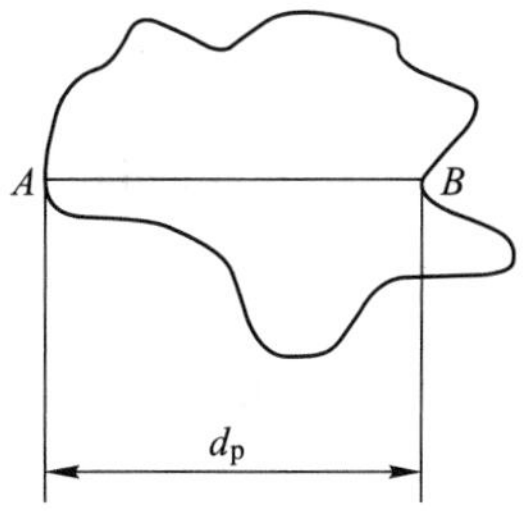

Fig. 1.2 Martin diameter of a particle

1.1.2.2 Volume Equivalent Diameter

For the volume equivalent diameter, a sphere with diameter d_{p_v} will have the same volume as the particle in question and will have the same mass if the densities are equal. This gives

$$d_{p_v} = (6V_p/\pi)1/3 \tag{1.1}$$

Where d_{p_v} ——the volume equivalent diameter of the particle;

V_p ——the particle volume.

1.1.2.3 Stokes Diameter

Stokes diameter is defined in terms of particle settling velocity. All particles with similar settling velocities are considered to be the same size, regardless of their actual size, composition or shape. When a sphere of the same density as the particle has the same settling velocity as that particle, the diameter of that particle is as same as the diameter of sphere. This diameter is called Stokes diameter. Stokes diameter can be calculated by

$$d_{p.\,Stokes} = \left[\frac{18\mu v_t}{C_c(\rho_p - \rho_g)g}\right]^{1/2} \tag{1.2}$$

Where $d_{p.\,Stokes}$ ——Stokes diameter of the particle;

μ ——the gas viscosity;

v_t ——the particle settling velocity;

ρ_p ——the particle density;

ρ_g ——gas density;

C_c ——the cunningham slip correction factor, which is given by

$$C_c = 1 + \frac{2\lambda}{d_p}\left[1.257 + 0.4\exp\left(-1.1\frac{d_p}{2\lambda}\right)\right] \tag{1.3}$$

Where λ ——the mean free path of gas, $\lambda = 0.0665\mu m$ (in standard conditions).

When the particle diameter d_p is larger than $1\mu m$, the correction factor C_c is nearly equal to 1. Since gas density ρ_g is much less than particle density ρ_p, equation (1.2) can be written as

$$d_{p.\,Stokes} = \left[\frac{18\mu v_t}{C_c\rho_p g}\right]^{1/2} \tag{1.4}$$

It is more reasonable to use Stokes diameter $d_{p.\,Stokes}$ to describe the particle motion behavior in gas stream because Stokes diameter $d_{p.\,Stokes}$ is derived from the settling velocity in air.

1.1.2.4 Aerodynamic Diameter

A characteristic particle diameter which is used extensively in aerosol science and technology is the equivalent aerodynamic diameter, which expresses particle size in a homogeneous manner. It is defined as the diameter of a sphere of density $1000kg/m^3$ and the same settling velocity as the particle under study. This means that particles of any shape or density will have the same aerodynamic diameter if their settling velocities are the same.

The aerodynamic diameter is useful because it can be correlated to the residence time of particles in the atmosphere and their deposition in the human respiration system. For example, the term of $PM_{2.5}$ means that the aerodynamic diameter of airborne particulate is less than 2.5μm. The concentration of $PM_{2.5}$ in the atmosphere is used to assess the air quality. And also, $PM_{2.5}$ particles are related to the possibility of penetrating to the lower parts of the human respiratory tract.

The aerodynamic diameter is defined by

$$d_{p_a} = d_{p_v}\left(\frac{\rho_p}{C_c\rho_0}\right)^{1/2} \tag{1.5}$$

Where d_{p_a} ——the aerodynamic diameter of the particle;

ρ_0 ——the density of a spherical particle, $\rho_0 = 1000\text{kg/m}^3$.

However, the volume equivalent diameter d_{p_v} is difficult to be determined. In practice, Martin diameter d_p can be used in equation (1.5) approximately. If Stokes diameter $d_{p.\text{Stokes}}$ is used, there is no correction factor C_c in (1.5). That is

$$d_{p_a} = d_{p.\text{Stokes}}(\rho_p/\rho_0)^{1/2} \tag{1.6}$$

Example 1.1 Compute the aerodynamic diameter of a spherical particle with diameter 5μm and with density 2000kg/m^3.

Solution

From equation (1.3), the correction factor C_c is

$$C_c = 1 + \frac{2\lambda}{d_p}\left[1.257 + 0.4\exp\left(-1.1\frac{d_p}{2\lambda}\right)\right] = 1.033$$

Using equation (1.5), we find the aerodynamic diameter is given by

$$d_{p_a} = d_{p_v}\left(\frac{\rho_p}{C_c\rho_0}\right)^{1/2} = 6.96\mu\text{m}$$

From this example, the aerodynamic diameter of a particle is larger than that of a spherical particle whose density is greater than 1000kg/m^3.

1.2 Size Distribution

So far we have considered a single particle. But in particulate pollutant problems we are concerned with groups of particles having a variety of sizes, as shown in Fig. 1.3. To discuss such groups of particles and to make useful calculations about their behavior in collection devices, we need some ways of describing the particle size distributions.

1.2.1 Particle Number Fraction

In order to establish the idea of the particle size distribution function, an example of a group of measurement data of fly ash from a coal boiler is given here. There are 1015 particles with the size ranges from 0 to 50μm, as listed as column 2 and column 3 in Table 1.1.

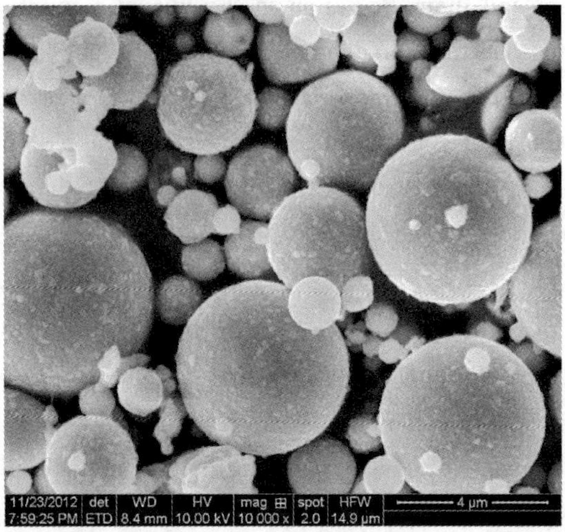

Fig. 1. 3　SEM image of a group of particles collected from a furnace that burns pulverized coal

Table 1. 1　Particle size distribution calculations

Serial i	Particle diameter $d_p/\mu m$	Number Δn_i	Fraction $\Delta F_i/\%$	Frequency $f=\Delta F_i/\Delta d_p$	Cumulative fraction $F/\%$
1	0~4	100	9. 8	0. 024	9. 8
2	4~6	160	15. 7	0. 079	25. 5
3	6~8	170	16. 7	0. 084	42. 2
4	8~9	70	6. 8	0. 069	49. 0
5	9~10	60	5. 9	0. 059	54. 9
6	10~14	190	18. 7	0. 047	73. 6
7	14~16	65	6. 4	0. 032	80. 0
8	16~20	80	7. 9	0. 02	87. 9
9	20~35	95	9. 5	0. 006	97. 4
10	35~50	25	2. 6	0. 001	100. 0
Total	—	$n_0=1015$	100	—	—

According to the particle number, a histogram can be drawn, as shown in Fig. 1. 4.

The fraction of particle number in each small size range is

$$\Delta F_i = \frac{\Delta n_i}{\sum \Delta n_i} = \frac{\Delta n_i}{n_0} \tag{1. 7}$$

The calculated values of the particle number fraction ΔF_i are listed in Table 1. 1.

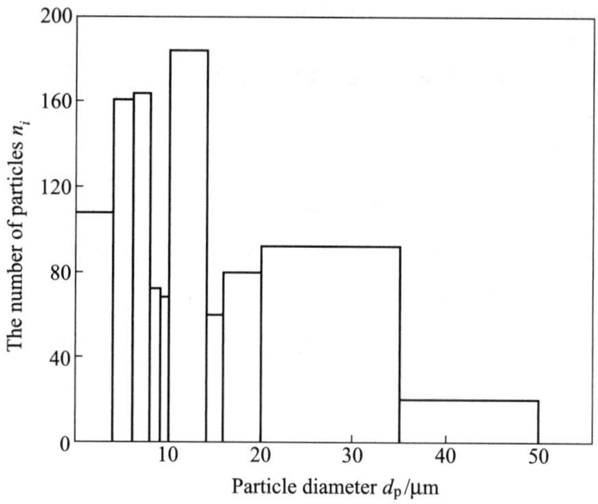

Fig. 1.4 Histogram of the particle number

1.2.2 Particle Frequency Distribution

Particle frequency is defined as

$$f(d_p)=\frac{\Delta F_i}{\Delta d_p}=\frac{\mathrm{d}F}{\mathrm{d}d_p} \tag{1.8}$$

The calculated values of the Particle frequency $\Delta F_i/\Delta d_p$ are also listed in Table 1.1. A histogram can be plotted in Fig. 1.5 according to the values of $\Delta F_i/\Delta d_p$.

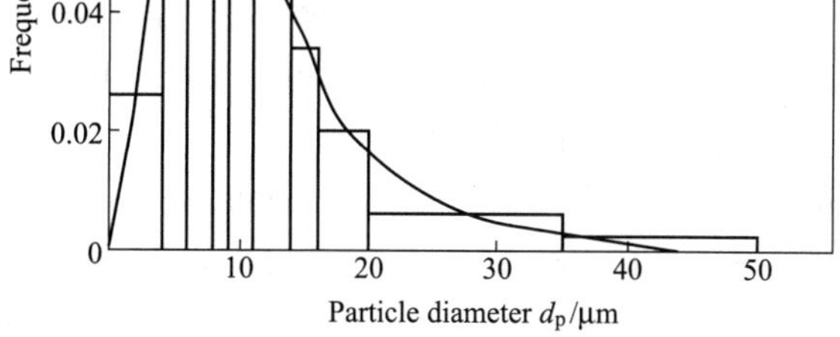

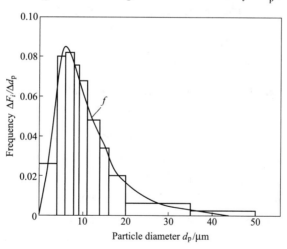

Fig. 1.5 Particle frequency distribution

If we connect the points which are located in the middle of the top side of each rectangular, a quite smooth curve of f is obtained. This curve of particle frequency distribution f follows the log-normal distribution, which is given by

$$f=\frac{\mathrm{d}F}{\mathrm{d}d_p}=\frac{1}{d_p\ln\sigma_g\sqrt{2\pi}}\exp\left[-\frac{(\ln d_p-\ln d_g)^2}{2(\ln\sigma_g)^2}\right] \tag{1.9}$$

Where d_g ——the geometric mean diameter of particulate;

σ_g ——the geometric standard deviation.

1.2.3　Particle Cumulative Frequency Distribution

The particle cumulative fraction is defined by

$$F = \sum_0^i \Delta F_i \tag{1.10}$$

This means that the percent by number is less than stated particle size. The results calculated by equation (1.10) are given in the last column in Table 1.1. If these values are connected together as shown in Fig. 1.6, the curve of F is also relatively smooth. This curve of F can also be expressed mathematically.

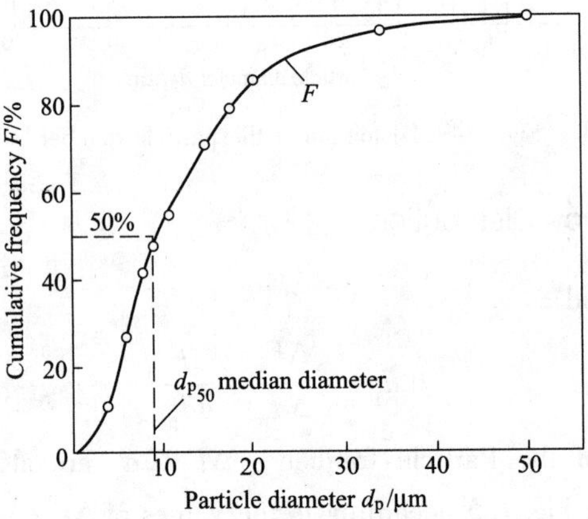

Fig. 1.6　Cumulative fraction of particles

From equation (1.8), we have

$$F = \int_0^{d_p} f \mathrm{d}d_p \tag{1.11}$$

Substituting equation (1.9) into equation (1.11), we find

$$F = \frac{1}{\ln\sigma_g \sqrt{2\pi}} \int_0^{d_p} \exp\left[- \frac{(\ln d_p - \ln d_g)^2}{2(\ln\sigma_g)^2} \right] \mathrm{d}(\ln d_p) \tag{1.12}$$

Equation (1.12) is called particle cumulative frequency distribution function. The curve of F in Fig. 1.6 can be described by equation (1.12).

Both frequency distribution equation (1.9) and cumulative frequency distribution equation (1.12) can be used to describe the particle size distributions. These two equations contain only two constants: d_g and σ_g. If these two values have been specified, the frequency distribution equation (1.9) and cumulative frequency distribution equation (1.12) have been specified.

For a group of n particles, geometric mean diameter is written as

$$d_g = (d_{p_1} \cdot d_{p_2} \cdots d_{p_n})^{1/n} \tag{1.13}$$

However, it is difficult to use equation (1.13) to determine the value of geometric mean diameter d_g. The value of d_g can be obtained by reading the point when the particle cumulative fraction F is 50% on Fig. 1.6. That is

$$d_g = d_{p_{50}} \tag{1.14}$$

Where $d_{p_{50}}$ ——the number median diameter.

Geometric standard deviation σ_g can be calculated by

$$\sigma_g = \frac{d_p(F = 84.1\%)}{d_{p_{50}}} = \frac{d_{p_{50}}}{d_p(F = 15.9\%)} \tag{1.15}$$

Where $d_p(F = 84.1\%)$ ——the particle diameter of which 84.1% of particle number less than this size;

$d_p(F = 15.9\%)$ ——the particle diameter of which 15.9% of particle number less than this size.

Even though the values of $d_p(F = 84.1\%)$ and $d_p(F = 15.9\%)$ can be found from Fig. 1.6, it is not easy to read the values precisely because the curve of F is not a straight line. Fortunately, if the data are log-normally distributed, the representation on a log-probability paper is a straight line.

For instance, if we put the data in the last column in Table 1.1 onto the log-probability paper, as illustrated in Fig. 1.7, the result is a straight line.

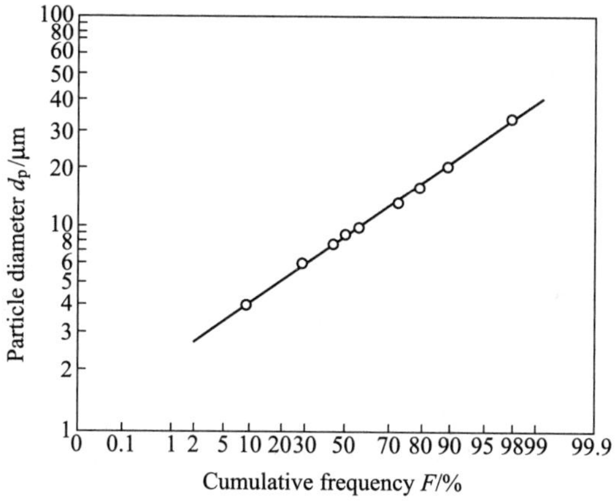

Fig. 1.7 Cumulative frequency distribution

From Fig. 1.7, $d_{p_{50}}$, $d_p(F = 84.1\%)$ and $d_p(F = 15.9\%)$ can be read directly, they are $d_{p_{50}} = 9\mu m$, $d_p(F = 84.1\%) = 18\mu m$, and $d_p(F = 15.9\%) = 4.5\mu m$.

Then, the geometric standard deviation σ_g is

$$\sigma_g = \frac{d_p(F = 84.1\%)}{d_{p_{50}}} = \frac{18\mu m}{9\mu m} = 2$$

or

$$\sigma_g = \frac{d_{p_{50}}}{d_p(F = 15.9\%)} = \frac{9\mu m}{4.5\mu m} = 2$$

Then, the frequency distribution function and cumulative frequency distribution function for fly ash in Table 1.1 are obtained respectively by

$$f = \frac{1}{d_p \ln 2\sqrt{2\pi}} \exp\left[-\frac{(\ln d_p - \ln 9)^2}{2(\ln 2)^2}\right]$$

$$F = \frac{1}{\ln 2\sqrt{2\pi}}\int_0^{d_p}\exp\left[-\frac{(\ln d_p - \ln 9)^2}{2(\ln 2)^2}\right]d(\ln d_p)$$

1.2.4 Particle Number and Mass Distribution

If we determine the distribution by catching the particles on a greased microscope slide and measuring the diameter of suitable number of particles, our results will be presented as the percent by number at various size range. Then, the particle distribution by number will be developed. However, in particulate pollutants separation, the particle motion behavior is related with the mass of particles. Furthermore, the concentration of particulate pollutants in gas is more often used as mass concentration. Therefore, the particle distribution by mass is the most common way of representing the data.

When we want to obtain the distribution of particle mass, the fraction of mass at various size ranges are given by

$$\Delta G_i = \frac{n_i d_{pi}^3}{\sum n_i d_{pi}^3} \tag{1.16}$$

The mass cumulative fraction is given by

$$G = \sum_0^i \Delta G_i \tag{1.17}$$

A logarithmic normal distribution can also be used to express the distribution of particle mass by

$$G = \frac{1}{\ln\sigma_g\sqrt{2\pi}}\int_0^{d_p}\exp\left[-\frac{(\ln d_p - \ln d_{p_{50}})^2}{2(\ln\sigma_g)^2}\right]d(\ln d_p) \tag{1.18}$$

Where $d_{p_{50}}$ ——mass median diameter;

σ_g ——mass geometric standard deviation.

Example 1.2 Measurement data of fine glass particles are listed in Table 1.2. Find the number median diameter, the mass median diameter, and the geometric standard deviation by means of a log-probability paper .

Table 1.2 Number cumulative fraction and mass cumulative fraction of a group of fine glass particles

Particle diameter d_p/μm	Number cumulative fraction F/%	Mass cumulative fraction G/%
5	0.0	0.0
10	1.0	0.1
15	13.8	0.6
20	42.0	10.5
25	68.0	28.5
30	85.0	50.0
35	93.0	67.9
40	97.2	80.8
45	98.8	89.2
50	99.5	94.0
55	99.8	97.0
60	99.9	98.1

Solution

The values of number cumulative fraction and mass cumulative fraction with respect to the particle diameters are drawn on the log-probability paper. Two straight lines are obtained when the data points are connected, shown in Fig. 1. 8. It has been found that these two lines are parallel.

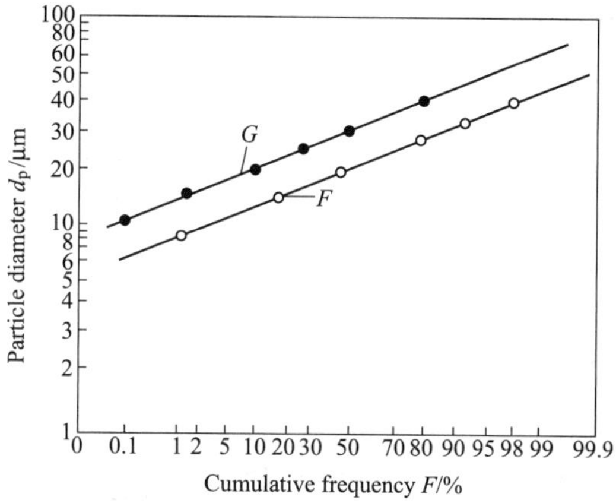

Fig. 1. 8 Number and mass cumulative frequency distributions of fine glass particles on log-probability paper

According to Fig. 1. 8, we find:

(1) The number median diameter is $d_{p_{50}} = 21.5\mu m$. The diameter of 84. 1% cumulative value by number is $d_p(F = 84.1\%) = 29.5\mu m$. The number geometric standard deviation is $\sigma_g = \frac{29.7}{21.5} = 1.39$.

(2) The mass median diameter is $d_{p_{50}} = 30\mu m$. The diameter of 84. 1% cumulative value by mass is $d_p(F = 84.1\%) = 41.7\mu m$. The mass geometric standard deviation is $\sigma_g = \frac{41.7}{30} = 1.39$.

It is clear that for one group of particles, the number median diameter is different from the mass median diameter. But the number geometric standard deviation is equal to the mass geometric standard deviation.

1. 3 Particle Concentration

1. 3. 1 Relation of Number and Mass Concentrations

There are two concentration definitions in airborne particulate separation. One is the number of particles divided by the volume of air. It is call as number concentration. The unit is m^{-3}. The number concentration is often used in particle nucleation, diffusion, agglomeration, particle deposition in the lungs, etc. The othefr is the mass of particulate divided by the volume of air. It is called as mass concentration. The units of mg/m^3, or $\mu g/m^3$ are commonly used for mass concentration. The mass concentration is often used in evaluation of the performance of the particle separators and the particulate emission from the pollutant source.

If the particle density ρ_p and particle frequency distribution f are known, the number concentration can be converted into the mass concentration.

Example 1.3 It is known an airborne particulate with density ρ_p follows the log-normal distribution. The number concentration of this particulate is n_0/m^3. What is the mass concentration?

Solution

Firstly, an increment of particle diameter Δd_p is taken from the particle number frequency distribution f, as shown in Fig. 1.9.

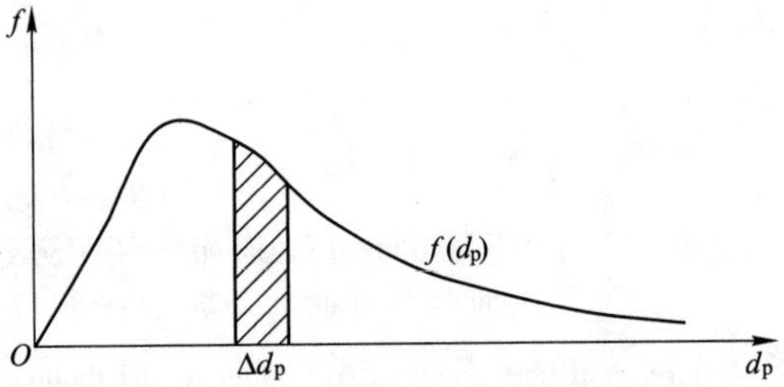

Fig. 1.9 Mathematic model of mass concentration development

According to eqnations (1.7) and (1.8), in the range of Δd_p, the number of particles with d_p is

$$\Delta n = n_0 \Delta F = n_0 f \Delta d_p \tag{1.19}$$

Here, in equation (1.19), the distribution function f is given by equation (1.9). Then, the volume of the particle in the range of Δd_p is

$$\Delta V = \frac{\pi}{6} d_p^3 n_0 f \Delta d_p \tag{1.20}$$

In the whole ranges of particle sizes, the total volume of the particulate per cubic meter can be obtain by integrating equation (1.20). It is writtten as

$$V = \frac{\pi}{6} n_0 \int_0^{\infty} \frac{1}{d_p \sqrt{2\pi} \ln\sigma_g} \exp\left(- \frac{\ln d_p - \ln d_{p_{50}}}{2\ln\sigma_g} \right) d_p^3 \mathrm{d}d_p \tag{1.21}$$

Let

$$\ln d_{p_{50}} = a \text{ , } \ln\sigma_g = \sigma \text{ , } \ln d_p = y \tag{1.22}$$

We have

$$d_p = e^y \text{ , } \mathrm{d}d_p = e^y \mathrm{d}y \tag{1.23}$$

From equation (1.22), when $d_p = 0$, then $y \to -\infty$, and when $d_p \to \infty$, then $y \to \infty$. Substituting equations (1.22) and (1.23) into equation (1.21), we have

$$V = \frac{\pi}{6} n_0 \int_{-\infty}^{\infty} \frac{1}{\sqrt{2\pi}\sigma} \exp\left[- \frac{(y-a)^2}{2\sigma^2} + 3y \right] \mathrm{d}y \tag{1.24}$$

Since

$$\left[- \frac{(y-a)^2}{2\sigma^2} + 3y \right] = \left[- \frac{1}{2\sigma^2}(y - a - 3\sigma^2)^2 + 3a + 4.5\sigma^2 \right]$$

Then, the solution of equation (1.24) is, obtained by

$$V = \frac{\pi}{6} n_0 e^{3a+4.5\sigma^2} \int_{-\infty}^{\infty} \frac{1}{\sqrt{2\pi}\sigma} \exp\left[-\frac{(y - a - 3\sigma^2)^2}{2\sigma^2} \right] dy = \frac{\pi}{6} n_0 e^{3a+4.5\sigma^2} \times 1 \quad (1.25)$$

Substituting equation (1.22) back to equation (1.25), we have

$$V = \frac{\pi}{6} n_0 \exp[3\ln d_{p_{50}} + 4.5(\ln \sigma_g)^2] \quad (1.26)$$

Finally, the mass concentration of this airborne particulate is

$$c = \rho_p \times V = \frac{\pi}{6} \rho_p n_0 \exp[3\ln d_{p_{50}} + 4.5(\ln \sigma_g)^2] \quad (1.27)$$

Therefore, the relation of the number concentration and the mass concentration is determined by equation (1.27).

1.3.2 Relation of Standards and Particulate Pollutant Concentration

The standards of particulate pollutant concentration belongs to air pollution control laws and regulations. The air pollution control laws and regulations are authorizations by local, state, or national legal authorities. The air pollutants consist of the gaseous pollutants and particulate pollutants. We only discuss the particulate pollutants in this book. Two standards of particulate pollutant concentration, the emission standard and the air quality standard, are commonly used in particulate pollutant control and pollution evaluation. The mass concentration is used in both emission standard and the air quality standard.

1.3.2.1 Emission Standard

The emission standard is the permitted upper limit concentration values of the pollutants emitted from the smoke stacks, or the discharging pipes. The basic idea of the emission standard is that there is some maximum possible (or practical) degree of emission control. This degree varies between various classes of emitters (e.g. cement factories, power plants, iron and steel companies). If this degree of control is determined for each class, every member of that class is required to limit emission to this maximum degree possible.

The emission standard is very important in particulate pollutant control because the emission concentration of the pollutant control equipment in a production system must satisfy the emission standard.

The emission standard is changeable. It is different from country to country and from time to time. It mainly depends on the development of economy, technology, and the consciousness of the environmental protection of citizens. As an example, Table 1.3 gives the comparison of particulate pollutant emission standards of kiln in cement factory. It is indicated that the standard for PM emission concentration in China approaches the standards in the developed countries.

Table 1.3 Comparison of particulate pollutant (PM) emission standards of kiln in cement factory (mg/m^3)

China (GB 4915—2013)	USA (NSPS)	Germany (TA Luft Guideline)	Japan (Air pollution prevention law)	UN (BAT)
General area 30	Old source 14	20	General area 100	10~20
Special area 20	New source 4		Special area 50	

The emission standard is also different in different industries. Power plant, iron and steel industry are the most typical emitters of the particulate pollutant. Here, the emission standards of particulate pollutant for thermal power plant and for iron and steel industry in China are extracted in Table 1.4.

Table 1.4 Emission standards of particulate pollutant for power plant and iron & steel industry (mg/m^3)

Thermal power plant Coal burning boiler (GB 13223—2011)	Iron and steel industry			
	Sintering (GB 28662—2012)	Iron smelt (GB 28663—2012)	Steel smelt (GB 28664—2012)	Steel rolling (GB 28665—2012)
Old project 30	Head of sinter 40	Blass furnace 15	Converter 50	Hot rolling 20
New project 20	Tail of sinter 20	Others 10	Others 15	Others 15

1.3.2.2 Air Quality Standard

Air quality standard is the maximum pollutant concentration permitted in air. It means that if the pollutant concentration in air is less than this standard value, there is possibly no harmful effects on human health, property, aesthetics, and the global climate. An air quality standard for particulate pollutant (PM) in China (GB 3095—2012) is shown in Table 1.5. It has been executed since 2016 in China.

Table 1.5 Air quality standard of particulate pollutant in China (GB 3095—2012)

Particulate	Time	Average concentration limit / $\mu g \cdot m^{-3}$		Remarks
		Area of class Ⅰ	Area of class Ⅱ	
TSP	One year	80	200	Area of class Ⅰ: natural protecting area, scenic spot, national park, etc. Area of class Ⅱ: city or town, residential area, cultural area, countryside, commercial district, industrial area, etc.
	24 hours	120	300	
PM10	Year	40	70	
	24 hours	50	150	
PM2.5	Year	15	35	
	24 hours	35	75	

1.3.3 Relation of Human Health and Particulate Pollutant

In health studies, particles in the size range 2.5~10.0μm (PM_{10}~$PM_{2.5}$) are referred to as the coarse fraction. Particles smaller than 2. 5 μm are referred to as the fine fraction and particles smaller than 0. 1 μm are termed the ultrafine fraction. It should be noted that PM_{10} actually also contains the $PM_{2.5}$ and UFPs, and likewise $PM_{2.5}$ also contains the ultrafine fraction. All these particulate-matter fractions in the atmosphere consist of aerosol.

The effects of the inhaled particles on human health are associated with size and concentration of the particulate pollutants, as well as the exposure time in the polluted environment[6,7].

1. 3. 3. 1 Deposition Positions of Inhaled Particle Sizes in Respiratory System

Airborne particulates (Aerosols) can cause health problems when deposited on the skin, but generally the most sensitive route of entry into the body is through the respiratory system. The deposition process is controlled by physical characteristics of the inhaled particles and the physiological factors of the individuals involved. Of the physical factors, particle size and size distribution are among the most important[8]. Fig. 1. 10 shows the deposited positions of the particle size ranges in the respiratory system. Therefore, sometimes, the particle size from 0. 1μm to 10μm is called respiratory particulate. It is indicated in Fig. 1. 10 that particle size from 0. 1μm to about 3μm is more harmful because the particulate of this size range can get into the lung. This is, obviously, one of the most important reasons to control $PM_{2.5}$.

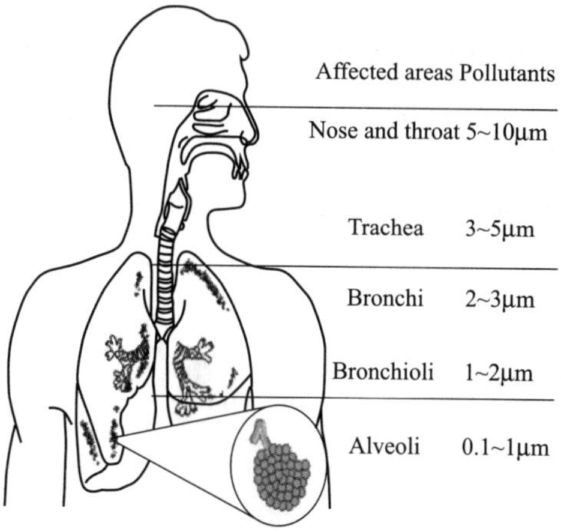

Fig. 1. 10 Deposition positions of inhaled particles with different size ranges in respiratory system

1. 3. 3. 2 Ultrafine Particle Effects

More and more recent researches have revealed that the ultrafine particles (UFP < 0. 1μm in aerodynamic diameter) can cause significant adverse health effects[9]. Report in their extensive review that the vast majority of the literature suggests that it is the fine and potentially also UFPs that are most associated with adverse health effects. The inhaled UFP can get into blood circulation[10] and brain of humans[11]. It has proved that the inhaled UFP may lead to ageing, cancer, and nervous sickness.

However, to investigate the health of effect of particulate pollutants is complex and difficult since the harmful dosage depends on the size distribution, concentration, toxicity, and the breathing time of airborne particulates. Furthermore, because there is currently no legislative requirement to measure UFP number concentrations in the ambient air, the number of health-related studies is somewhat limited, with none available for large population groups, particularly compared to the number of studies available in relation to both $PM_{2.5}$ and PM_{10}.

Exercises

1.1 A spherical particle has a density of 2×10^3 kg/m³, and the terminal settling velocity of this particle in still gas is 0.01m/s. When the gas density is 1.2kg/m³ and the gas viscosity is 1.85×10^{-5} Pa · s, calculate the particle size.

1.2 Particulate matter from operation is known to have a log-normal distribution. The data is shown in Table 1.6.

Table 1.6 Particle size analysis

Particle size/μm	Total weight in each fraction/%	Cumulative/%
0~10	39.6	36.9
10~20	19.1	56.0
20~40	18.0	74.0
>40	26.0	100.0

(1) Determine the geometric mean diameter and geometric deviation.

(2) Plot a frequency distribution curve.

1.3 Sketch the frequency distribution curve according to Fig. 1.11 and Fig. 1.12.

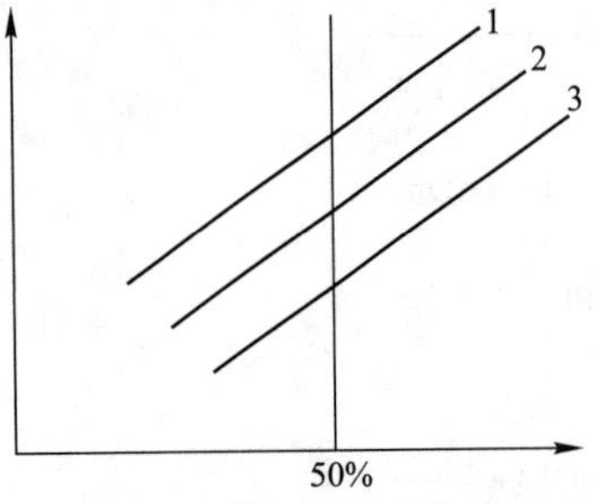

Fig. 1.11 Log-probability curve

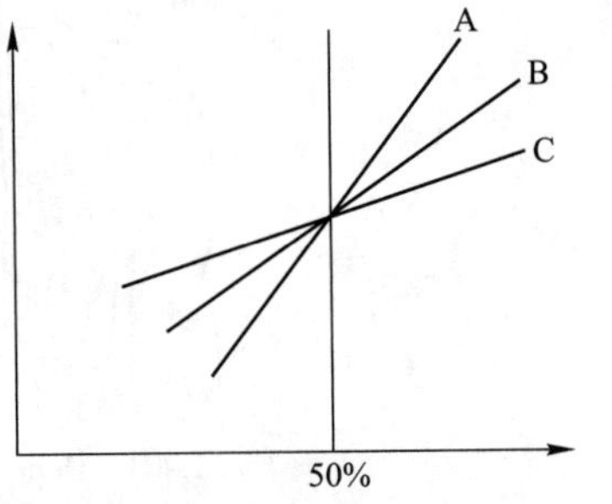

Fig. 1.12 Log-probability curve

1.4 The number concentration of the log-normal distribution airborne particulate is 10^8/m³. The number median diameter is 5μm, the geometric deviation is 1.5, and the density is 2×10^3 kg/m³. Calculate

(1) mass median diameter,

(2) mass concentration.

1.5 A group of particles is described by the log-normal distribution with mass median diameter d_{P50} = 5μm and geometric deviation σ_g = 1.2.

(1) What fraction by mass of the particles have diameters less than 1μm?

(2) What fraction by number of the particles have diameters less than 1μm?

1.6 From the line on Fig. 1.8, estimate the diameter that corresponds to 10% by mass and number respectively.

1.7 If a population of particles is log-normal distribution with mass median diameter $d_{p_{50}} = 10\mu m$ and geometric deviation $\sigma_g = 1.5$, what is the diameter that has 99% of mass smaller than it? And what is the diameter that has 1% of the mass smaller than it?

1.8 Describe the difference between the particle emission standard and air quality standard.

References

[1] Heal M R, Kumar P, Harrison R M, et al. Particles, air quality, policy and health [J]. Chemical Society Reviews, 2012, 41 (19): 6606-6630.

[2] Cheng Z, Wang S, Jiang J, et al. Long-term trend of haze pollution and impact of particulate matter in the Yangtze River Delta, China [J]. Environmental Pollution, 2013, 182: 101-110.

[3] Colbeck I, Mihalis L. *Aerosol Science Technology and Applications* [M]. West Sussex, United Kingdom, John Wiley & Sons Ltd., 2014.

[4] Moser P, Schmidt S, Stahl K, et al. The wet electrostatic precipitator as a cause of mist formation—Results from the amine-based post-combustion capture pilot plant at Niederaussem [J]. International Journal of Greenhouse Gas Control, 2015, 41: 229-238.

[5] Martin C. *Air Pollution Control Theory* [M]. New York, Mcgraw-Hill, 1976.

[6] Pope C A, Dockery D W. Health effects of fine particulate air pollution: Lines that connect [J]. Journal of the Air & Waste Management Association, 2006, 56 (10): 709-742.

[7] Valavanidis A, Fiotakis K, Vlachogianni T, et al. Airborne particulate matter and human health: toxicological assessment and importance of size and composition of particles for oxidative damage and carcinogenic mechanisms [J]. Journal of Environmental Science and Health Part C-environmental Carcinogenesis & Ecotoxicology Reviews, 2008, 26 (4): 339-362.

[8] Hofmann W. Modelling inhaled particle deposition in the human lung—A review [J]. Journal of Aerosol Science, 2011, 42 (10): 693-724.

[9] Gurgueira S A, Lawrence J, Coull B A, et al. Rapid increases in the steady-state concentration of reactive oxygen species in the lungs and heart after particulate air pollution inhalation [J]. Environmental Health Perspectives, 2002, 110 (8): 749-755.

[10] Nemmar A, Hoet P, Vanquickenborne B, et al. Passage of inhaled particles into the blood circulation in humans [J]. Circulation, 2002, 105 (4): 411-414.

[11] Oberdorster G, Sharp Z D, Atudorei V, et al. Translocation of inhaled ultrafine particles to the Brain [J]. Inhalation Toxicology, 2004, 16 (67): 437-445.

2 Motion of Aerosol Particulate in an External Force Field

2.1 Drag Force on a Single Spherical Particle

A good place to start to study the dynamical behavior of particles in gas is to consider the drag force exerted on a particle as it moves in gas. This drag force will always be present as long as the particle is moving[1].

The drag force is given in terms of the projected frontal area A_p, the relative velocity between the particle and gas v, and the drag coefficient C_D, defined by equation (2.1) as

$$f = C_D A_p \frac{\rho v^2}{2} \tag{2.1}$$

For a spherical particle, $A_p = \frac{\pi d_p{}^2}{4}$. Then, equation (2.1) becomes

$$f = C_D \frac{\pi d_p{}^2}{4} \frac{\rho v^2}{2} \tag{2.2}$$

Where d_p ——Martin or Stokes diameter.

In later discussion, we just use d_p to represent the particle diameter no matter what kind of particle diameter is.

It is clear that if the drag coefficient is determined, the drag force on the particle can be calculated. It had been found by fluid mechanics and experiments that the drag coefficient C_D is a function of the Reynolds number Re, as shown in Fig. 2.1. The Reynolds number is defined by

$$\mathrm{Re} = \frac{\rho d_p v}{\mu} \tag{2.3}$$

Where μ——the kinematic viscosity of the gas in Pa · s.

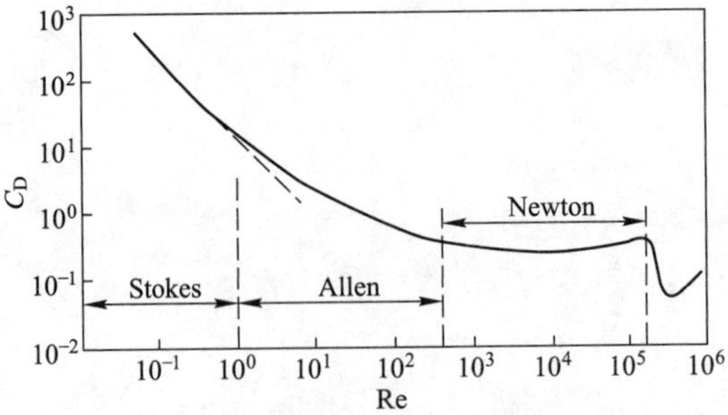

Fig. 2.1 Relation of drag coefficient C_D and Reynolds number Re

According to the Reynolds number Re, the flow state can be divided into three ranges, including laminar regime, transition regime, and turbulent regime[2].

2.1.1 Laminar Regime

The net force exerted by the fluid on the moving particle may be derived from the solution of the Navier-Stokes equations[3], which describe fluid motion. The exerted force is obtained by integrating the normal and tangential forces over the particle surface. In addition, for small particles, momentum transfer (from the fluid to the particle) does not occur in the continuum momentum transfer regime, leading to a decrease of the Stokes drag force. The total resisting force acting on a small spherical particle of velocity v in the fluid is expressed as

$$f = 3\pi\mu d_p \nu \quad \mathrm{Re} < 0.1 \tag{2.4}$$

Equation (2.4) is called Stokes' law. Let equation (2.4) be equal to equation (2.2), and reference equation (2.3), the drag coefficient is

$$C_D = 24/\mathrm{Re} \tag{2.5}$$

When Re = 1, the value calculated by equation (2.4) is 1.3% less than the real value. A more accurate drag coefficient formula had been given by Oseen when the inertial parts in Navier-Stokes equations was considered by

$$C_D = \frac{24}{\mathrm{Re}}\left(1 + \frac{3}{16}\mathrm{Re} + \frac{9}{160}\mathrm{Re}^2 \ln 2\mathrm{Re}\right) \quad 0.1 < \mathrm{Re} \leqslant 2 \tag{2.6}$$

Even though the Oseen formula is more precise theoretically, the inaccuracy of stokes law is very small. For the approximate calculation, the Reynolds number can be extend $\mathrm{Re} \leqslant 1$ for application of equation (2.4).

2.1.2 Transition Regime

Transition regime is located between laminar and turbulent regime. The Reynolds number is about $1<\mathrm{Re}<500$. In transition regime, the drag coefficient has been given by Allen as

$$C_D = 10.6/\sqrt{\mathrm{Re}} \quad 1 < \mathrm{Re} \leqslant 500 \tag{2.7}$$

Therefore, the transition regime is also called the Allen regime.

2.1.3 Turbulent Regime

Turbulent regime is also called Newton regime. The drag coefficient is

$$C_D = 0.44 \quad 500<\mathrm{Re}<2\times10^5 \tag{2.8}$$

In the practice of the particulate pollutant control, the flow around a moving particle is usually in the Stokes regime ($\mathrm{Re} < 1$) because the particle is quite small, and the relative velocity between particle and gas stream is low. For example, in an electrostatic precipitator, the gas velocity in the electric field is about 1m/s, while in the fiber filtration, the filtration velocity is only about 0.02m/s. Therefore, the drag force act upon the particles usually satisfy Stokes' law in the most cases.

2. 1. 4 Particles too Small for Stokes' Law

When the particle becomes very small, another of the assumptions leading to Stokes' law becomes inaccurate. Stokes' law assumes that the fluid in which the particle is moving is a continuous medium. However, real gases are not truly continuous but made up of atoms and molecules. As long as the particle we are considering is much larger than the space between the individual gas molecules or atoms, the fluid interacts with the particle as if it were a continuous medium. When a particle becomes as small as or smaller than the average distance between molecules, its interaction with molecules changes. The effect of the change is to lower the drag force, which causes the particle to move faster. The most widely used correction factor for this change has the form as

$$f = 3\pi\mu d_{p}/C_{c} \tag{2.9}$$

Where C_c ——Cunningham slip correction factor, given by equation (1. 3).

2. 2 Motion of a Particle in an External Force Field

The motion of a particle arises in the first place due to the action of some external force on the particle, such as gravity or electrical forces. The drag force appears as soon as there is a difference between the velocity of the particle and that of the fluid. For the variety of applications to design of particle collection devices, it is necessary to describe the motion of a particle in either a quiescent or flowing fluid subject to external forces on a particle. The basis of this description is an equation of motion for a particle.

Summing forces acting on the spherical particle and equating this sum to the rate of change of momentum in accordance with Newton' s second law of motion, gives

$$m_{p}\frac{d\vec{v}}{dt} = \frac{3\pi\mu d_{p}}{C_{c}}(\vec{u} - \vec{v}) + \sum_{i}\vec{F}_{i} \tag{2.10}$$

Where m_p ——the mass of the particle;

$\vec{u}$ ——the gas velocity;

$\vec{v}$ ——the particle velocity;

$\sum_{i}\vec{F}_{i}$ ——the sum of external forces.

2. 2. 1 Motion of a Particle in Gas under Gravity

2. 2. 1. 1 Basic Motion Equation

Consider for the motion of a particle in a gas in the presence only of gravity. The buoyant force of the particle in the gas is neglected because the density of gas is much smaller than that of a solid particle. The equation (2. 10) becomes

$$m_{p}\frac{d\vec{v}}{dt} = \frac{3\pi\mu d_{p}}{C_{c}}(\vec{u} - \vec{v}) + m_{p}g \tag{2.11}$$

If we divide the equation by $3\pi\mu d_p/C_c$, and notice $m_p = (\pi/6)d_p^3\rho_p$, we obtain

$$\tau\frac{d\vec{v}}{dt} + \vec{v} = \vec{u} + \tau g \tag{2.12}$$

Where τ ——relaxation time, given by

$$\tau = \frac{\rho_p d_p^2 C_c}{18\mu} \tag{2.13}$$

As an example of equation (2.12), consider the motion of a particle in the x, z-plane, where the z-axis is taken positive downward, as shown in Fig. 2.2. Then, equation (2.12) becomes

$$\tau\frac{dv_x}{dt} + v_x = u_x \tag{2.14}$$

$$\tau\frac{dv_z}{dt} + v_z = u_z + \tau g \tag{2.15}$$

Equations (2.14) and (2.15) can be written as

$$\tau\frac{d^2x}{dt^2} + \frac{dx}{dt} = u_x \tag{2.16}$$

$$\tau\frac{d^2z}{dt^2} + \frac{dz}{dt} = u_z + \tau g \tag{2.17}$$

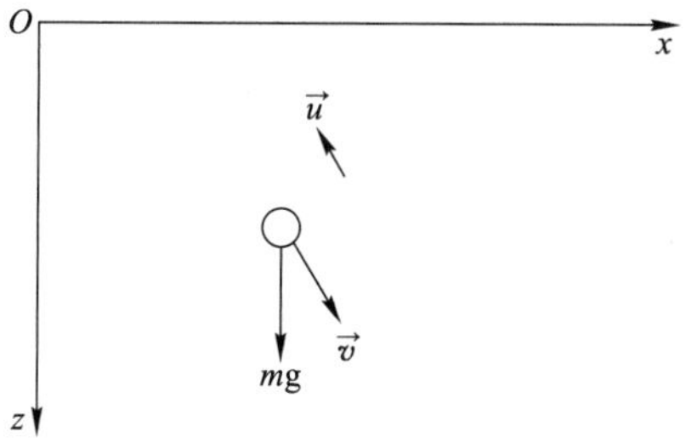

Fig. 2.2 Motion of a particle under gravity

If at time zero, $t=0$, the initial velocity components of the particle in the two directions x and y are $v_x(0) = v_{x_0}$ and $v_z(0) = v_{z_0}$, and if the gas velocity components are constant, we can integrate equations (2.16) and (2.17) to obtain

$$v_x(t) = u_x + (v_{x0} - u_x)e^{-t/\tau} \tag{2.18}$$

$$v_z(t) = (u_z + \tau g) + (v_{z0} - u_z - \tau g)e^{-t/\tau} \tag{2.19}$$

If the particle starts, for example, at the origin, its position at any time is found by integrating equations (2.18) and (2.19) once more or by solving equations (2.16) and (2.17) directly with initial conditions $x(0) = z(0) = 0$ and $(dx/dt)_0 = v_{x_0}$, $(dz/dt)_0 = v_{z_0}$, to obtain

$$x(t) = u_x t + \tau(v_{x_0} - u_x)(1 - e^{-t/\tau}) \tag{2.20}$$

$$z(t) = (u_z + \tau g)t + \tau(v_{z_0} - u_z - \tau g)(1 - e^{-t/\tau}) \tag{2.21}$$

2.2.1.2 Terminal settling velocity

If the particle is at rest at $t=0$ and the air is still, that is $v_{z_0} = 0$ and $u_z = 0$, the only velocity component is the z-direction, from equation (2.19), we have

$$v_z(t) = \tau g(1 - e^{-t/\tau}) \tag{2.22}$$

For $t \gg \tau$, the particle attains a constant velocity, called its terminal settling velocity as

$$v_t = \tau g = C_c \frac{\rho_p d_p^2}{18\mu} g \tag{2.23}$$

It can be seen that τ is the characteristic time for the particle to approach steady motion. Similarly if a particle enters a moving airstream, it approaches the velocity of the stream with a characteristic time τ (relaxation time). Values of τ and terminal settling velocity for unit density spheres at 298K in air are shown in Table 2.1.

Table 2.1 Relaxation time and terminal settling velocity of spheres with density of 1×10^3kg/m^3 at 298K in air

d_p/μm	τ/s	v_t/cm · s^{-1}
0.1	4.0×10^{-8}	8.8×10^{-5}
0.5	9.0×10^{-8}	1.0×10^{-3}
1.0	3.6×10^{-6}	3.5×10^{-3}
5.0	8.0×10^{-5}	7.8×10^{-2}
10.0	3.1×10^{-4}	3.1×10^{-1}

2.2.1.3 Stokes Stopping Distance

If a particle starts at v_{x_0} initially in still air, in the x direction, form equation (2.20), the distance at any time of t is

$$x(t) = \tau v_{x_0}(1 - e^{-t/\tau}) \tag{2.24}$$

When the particle velocity is getting to zero, the distance of the particle movement is called stopping distance. In equation (2.24), let $t \to \infty$, then the stopping distance x_s is

$$x_s = \tau v_{x_0} \tag{2.25}$$

Example 2.1 A spherical particle with size of 10μm and density of 3×10^3kg/m^3 is projected upward at velocity 5m/s into sill air at 298K. Determine the highest vertical distance of the particle movement and the time.

Solution

Because the particle size is much larger than 1μm, and Cunningham slip correction factor $C_c \approx 1$. The particle has only vertical motion, and air is still, that is $u_z = 0$. From equation (2.19), we have

$$v_z(t) = \tau g + (v_{z_0} - \tau g)e^{-t/\tau} \tag{2.26}$$

Relaxation time is

$$\tau = \frac{\rho_p d_p^2}{18\mu} = \frac{3\times10^3\times(10\times10^{-6})^2}{18\times1.85\times10^{-5}} = 9\times10^{-4}\ (\text{s})$$

When the particle gets up to the highest point, $v_z = 0$. When $t=0$, $v_0=-5\text{m/s}$ (negative means the particle moving direction opposites to the z-direction). According to equation (2.26), the time of particle going to the highest point is

$$t = -\tau \ln \frac{\tau g}{5 + \tau g} = -9 \times 10^{-4} \ln \frac{9 \times 10^{-4} \times 9.81}{5 + 9 \times 10^{-4} \times 9.81} = 0.0057\ (\text{s})$$

From equation (2.21), the distance of the particle rising is

$$z(t) = \tau g t + \tau(v_{z_0} - \tau g)(1 - e^{-t/\tau}) \approx \tau g t + \tau(v_{z_0} - \tau g) \quad (2.27)$$

Then, the distance of the particle movement after $t = 0.0057\text{s}$ is

$$\begin{aligned} z &\approx \tau g t + \tau(v_{z_0} - \tau g) = 9 \times 10^{-4} \times 9.81 \times 0.0057 + \\ &\quad 9 \times 10^{-4}(-5 - 9 \times 10^{-4} \times 9.81) \\ &\approx -0.45 \times 10^{-2}\ (\text{m}) \end{aligned}$$

If this particle is projected in to air at 5m/s horizontally, according to the stopping distance equation (2.25), we find

$$x_s = \tau v_{x_0} = 9 \times 10^{-4} \times 5 = 0.45 \times 10^{-2}\ (\text{m})$$

It is indicated that the stopping distance is the same no matter what moving direction of a particle.

2.2.2 Motion of a Particle under a Centrifugal Force

When a particle moves along a circumference with radius r, a centrifugal force is presented as

$$F_c = \frac{\pi}{6} \rho_p d_p^3 \frac{u^2}{r} \quad (2.28)$$

Where u——the tangential velocity of the rotation flow, as shown in Fig. 2.3.

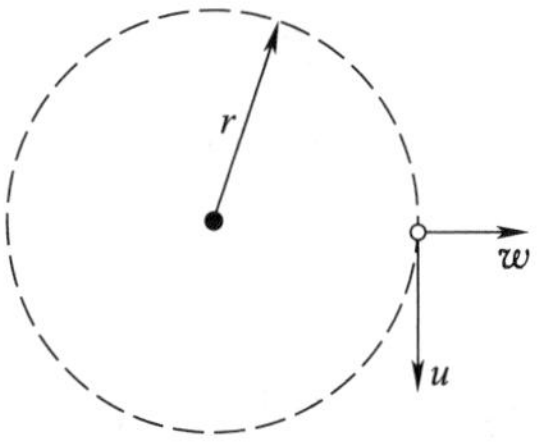

Fig. 2.3 Motion of a particle under a centrifugal force

The equation of a force balance in the radial direction is

$$F_c - f = m_p \frac{dw}{dt} \quad (2.29)$$

Where w —— centrifugal settling velocity of the particle in radial direction.

From above discussion, the terminal settling velocity of the fine particle gets to constant very quickly. Thus, $dw/dt = 0$, equation (2.29) becomes

$$\frac{\pi}{6} \rho_p d_p^3 \frac{u^2}{r} - 3\pi\mu d_p w / C_c = 0 \quad (2.30)$$

Solving the equation (2.30), the centrifugal settling velocity of the particle is obtained by

$$w = C_c \tau \frac{u^2}{r} \tag{2.31}$$

2.2.3 Motion of a Charged Particle in an Electric Field

It is important to discuss the motion of a charged particle in an electric field for particulate separation and particle measurement.

If a particle with charge q presents in a static electric field with electric field strength, E, the electrostatic force acting on the particle is $F = qE$. The equation of motion for a particle of charge q moving at velocity ω in the presence of electric field of strength, E, is given by

$$m_p \frac{d\omega}{dt} = Eq - \frac{3\pi\mu d_p}{C_c}\omega \tag{2.32}$$

Where q——particle charge;

E——the electric field strength.

At steady state in a stream, the electric force is balanced by the drag force as

$$\frac{3\pi\mu d_p}{C_c}\omega = Eq \tag{2.33}$$

The velocity of a particle in an electric field, called the migration velocity[4], is

$$\omega = EqC_c/3\pi\mu d_p \tag{2.34}$$

or

$$\omega = B_e E \tag{2.35}$$

Where B_e——the electrical mobility, given by

$$B_e = qC_c/3\pi\mu d_p \tag{2.36}$$

If a particle with charge q presents in an alternating electric field, the electric field strength is

$$E = E_0 \sin\psi t \tag{2.37}$$

Where E_0——the maximum electric field strength;

ψ——the radian frequency;

t——the time.

The electric force acted on the charged particle is

$$F = qE_0 \sin\psi t = F_0 \sin\psi t \tag{2.38}$$

Where F_0——the maximum electric force.

When the drag force on the charged particle follows Stokes law, the motion equation is given by

$$\frac{dv}{dt} + \frac{v}{\tau} - \frac{F_0}{m_p}\sin\psi t = 0 \tag{2.39}$$

Where m_p——the mass of the particle;

v——the particle velocity.

At the initial conditions of $t = 0$ and $v = 0$, the solution of the differential equation (2.39) is

$$v = \frac{\tau F_0}{m_p\sqrt{1+\tau^2\psi^2}}[\sin(\psi t - \varphi) + \exp(-t/\tau)\sin\varphi] \tag{2.40}$$

Where

$$\sin\varphi = \frac{\tau\psi}{\sqrt{1+\tau^2\psi^2}},\ \cos\varphi = \frac{1}{\sqrt{1+\tau^2\psi^2}} \tag{2.41}$$

In the right side of equation (2.40), $\exp(-t/\tau)\sin\varphi$ is going to zero very quickly. Then, equation (2.40) becomes

$$v = \frac{\tau q E_0}{m_p}\cos\varphi\sin(\psi t - \varphi) \tag{2.42}$$

From equation (2.41), we obtain

$$\tan\varphi = \psi\tau = 2\pi\tau/T \tag{2.43}$$

Where T——the period of the alternating electric field.

For a small particle, τ/T is small either. Then, $\tan\varphi \to 0$. Thus, equation (4.42) becomes

$$v = \frac{\tau q E_0}{m_p}\sin\psi t \tag{2.44}$$

2.3 Suspension of a Particle in a Duct

The suspension of a particle is a reverse problem of the deposition. Particle suspension occurs when aerodynamic lift forces become greater than the adhesive forces and the particle gravitational force. To discuss the suspension behavior of particles is useful in the aerodynamic transportation of particulate and the dust re-entrainment from collection surface.

When a particle deposits on a plane wall, this particle can be blown a float in air if a parallel gas stream is flowing along the wall, as shown in Fig. 2.4.

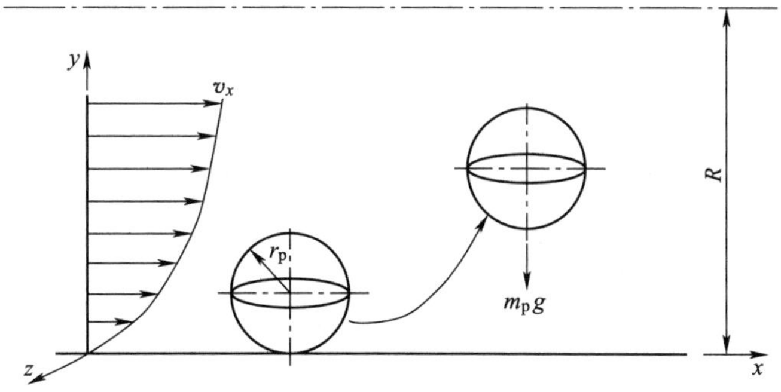

Fig. 2.4 Suspension of a particle on plane wall parallel to gas flow

The lift forces acting on the particles deposited on or near the wall in gas stream include wall-bounded shear force and the force of energy transfer from the turbulent fluid to a round particle.

Since the mechanism of energy transfer left force is complex, no analytical solution has been available till now. Here, as one of main energy transfer left forces caused by the difference of pressure between the top side and bottom side of a particle is discussed.

2.3.1 Particle Suspension Caused by a Shear Lift Force

A shear lift force will act on a particle due to the gas velocity distribution in the duct. The shear lift force L_s is given by Mei[5] as

$$L_s = C_s \frac{\pi d_p^2}{4} \frac{\rho v^2}{2} \tag{2.45}$$

Where ρ ——the gas density;

v ——the mean velocity in main flow area;

C_s ——the lift coefficient, which is determined by Reynolds number [equation (2.3)] of the flow around the particle, as shown in Fig. 2.5[5].

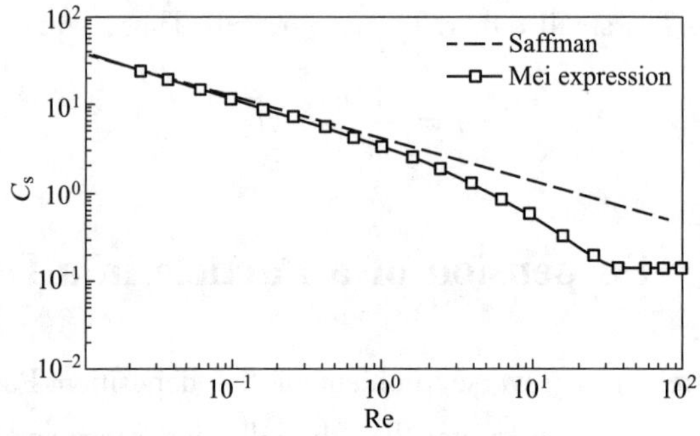

Fig. 2.5　Shear lift coefficient

(From Mei, 1992)

It is noticed that when Reynolds number is greater than 30, C_s is keeping a constant, $C_s \approx 0.15$. If a particle can be kept in suspension, the shear lift force duct must be satisfied by

$$L_s \geqslant m_p g \tag{2.46}$$

2.3.2 Particle Suspension Caused by Pressure Difference

The pressure of the top side and bottom side of a particle is different because the gas velocity in the boundary layer is not uniform. This pressure difference will lead to the particle suspension. It is called the lift force of pressure difference.

Suppose the static pressure at any point on the surface of the sphericalparticle is p_s, and the pressure at the point far away from the particle in the gas flow is p_∞. According to Bernoulli Equation, given by

$$p_s - p_\infty = \frac{1}{2}\rho v_x^2 \tag{2.47}$$

Assuming the maximum velocity at the axes of the duct is v_{max} (the distance from the wall R as shown in Fig. 2.4), and the velocity distribution of the flow section is[2]

$$v_x = \left(\frac{y}{R}\right)^{1/n} v_{max} \tag{2.48}$$

where, n is determined by Reynolds number, given in Table 2.2.

Table 2.2 Relation of *n* in equation (2.48) and Reynolds number Re of turbulent flow in a duct

Ranges of Re	$5\times10^3\sim5\times10^4$	$5\times10^4\sim5\times10^5$	$5\times10^5\sim1\times10^6$	$1\times10^6\sim5\times10^6$
n	6	7	8	9

In the industrial particulate pollutant control system, the Reynolds number of the gas flow is the range usually from 10^4 to 5×10^5. For instance, the velocity is $1\sim10$m/s in a duct with diameter of $0.5\sim1$m, the Reynolds number is given by

$$\mathrm{Re}=\frac{\rho dv}{\mu}=\frac{1.2\times(0.5\sim1)\times(1\sim10)}{1.85\times10^{-5}}=3\times10^4\sim6\times10^5$$

According to Table 2.2, $n=7$. Then equation (2.48) becomes

$$v_x=\left(\frac{y}{R}\right)^{1/7}v_{\max} \tag{2.49}$$

For the flow in a rectangular duct, the relation of average velocity v defined by the maximum velocity $v_{\max}$ is given by

$$v=\frac{1}{R}\int_0^R\left(\frac{y}{R}\right)^{1/7}v_{\max}\mathrm{d}y=\frac{7}{8}v_{\max} \tag{2.50}$$

For the flow in a cylindrical tube with radius of R, the velocity distribution of turbulent flow is

$$v_x=\left(1-\frac{r}{R}\right)^{1/7}v_{\max} \tag{2.51}$$

The relation of v defined by $v_{\max}$ is given by

$$v=\frac{1}{\pi R^2}\int_0^R\left(1-\frac{r}{R}\right)^{1/7}v_{\max}\times2\pi r\mathrm{d}r=2v_{\max}\int_0^R\left(1-\frac{r}{R}\right)^{1/7}\left(\frac{r}{R}\right)\mathrm{d}\left(\frac{r}{R}\right) \tag{2.52}$$

Let $(r/R)=t$, equation (2.52) becomes

$$v=2v_{\max}\int_0^1(1-t)^{1/7}t\mathrm{d}t \tag{2.53}$$

In equation (2.53), since $t\leqslant1$, $(1-t)^{1/n}$ can be converted to series, written as

$$(1-t)^{1/7}=1-\frac{1}{n}t+\frac{\frac{1}{7}\left(\frac{1}{7}-1\right)}{2!}t^2-\cdots\approx1-\frac{1}{7}t \tag{2.54}$$

Then, the average velocity in a duct for the turbulent flow is

$$v\approx2v_{\max}\int_0^1\left(1-\frac{t}{7}\right)t\mathrm{d}t=2v_{\max}\int_0^1\left(t-\frac{t^2}{7}\right)\mathrm{d}t=\frac{19}{21}v_{\max} \tag{2.55}$$

Take an cylindrical element from a particle as shown in Fig. 2.6. Suppose the range of the Reynolds number is 10^4 to 5×10^5. The velocity near any point on the surface of the spherical particle can be given from equation (2.51), written as

$$v_x=\left(\frac{r_{\mathrm{p}}+r_{\mathrm{p}}\sin\beta}{R}\right)^{1/7}v_{\max} \tag{2.56}$$

The area of the cylindrical element is $\mathrm{d}A=2\pi r_{\mathrm{p}}\cos\beta\times r_{\mathrm{p}}\mathrm{d}\beta$. Then overall lift force of the spherical particle caused by pressure difference is

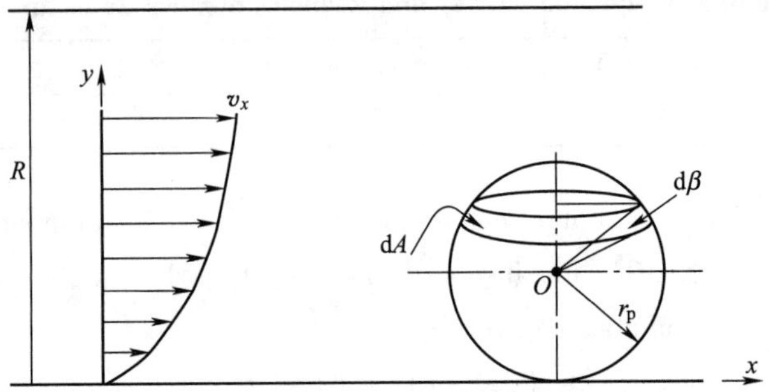

Fig. 2. 6　An element of a particle in analysis of the lift force caused by pressure difference

$$L_p = \int_A (p_s - p_\infty)\sin\beta dA = \int_{-\pi/2}^{\pi/2} \frac{1}{2}\rho v_x^2 (2\pi r_p^2 \cos\beta\sin\beta)d\beta \tag{2.57}$$

Substituting equation (2. 56) into equation (2. 57), we have

$$L_p = \pi\rho r_p^2 v_{max}{}^2 \left(\frac{r_p}{R}\right)^{2/7} \int_{-\pi/2}^{\pi/2} (\sin\beta + 1)^{2/7}\sin\beta\cos\beta d\beta \tag{2.58}$$

Let $t = \sin\beta$, equation (2. 58) becomes

$$L_p = \pi\rho r_p^2 v_{max}{}^2 \left(\frac{r_p}{R}\right)^{2/7} \int_{-1}^{1} (t + 1)^{2/7} t dt \tag{2.59}$$

It is not easy to solve the integral in equation (2. 59). It is noticed that $t \leqslant 1$. The brackets term in the integral can be also converted to series approximately as

$$(t + 1)^{2/7} = 1 + \frac{2}{7}t + \frac{\frac{2}{7}\left(\frac{2}{7} - 1\right)}{2!}t^2 + \cdots \approx 1 + \frac{2}{7}t - \frac{5}{49}t^2 \tag{2.60}$$

Therefore, the lift force on the spherical particle caused by pressure difference is given by

$$\begin{aligned} L_p &= \pi\rho r_p^2 v_{max}{}^2 \left(\frac{r_p}{R}\right)^{2/7} \int_{-1}^{1} (t + 1)^{2/7} t dt \\ &\approx \pi\rho r_p^2 v_{max}{}^2 \left(\frac{r_p}{R}\right)^{2/7} \int_{-1}^{1} \left(1 + \frac{2}{7}t - \frac{5}{49}t^2\right) t dt \\ &= \frac{4}{21}\pi\rho r_p^2 \left(\frac{r_p}{R}\right)^{2/7} v_{max}^2 \\ &= \frac{1}{21}\pi\rho d_p^2 \left(\frac{d_p}{2R}\right)^{2/7} v_{max}^2 \end{aligned} \tag{2.61}$$

If the particle on the wall is suspended (The adhesion force of Van der Waals between the particle and the wall is neglected), it must be

$$L_p \geqslant m_p g \tag{2.62}$$

There are two explanations for the theoretical model of a particle suspension caused by pressure difference: (1) the adhesive forces are not considered in equation (2. 62), and (2) if a particle has already floated in the gas, equation (2. 62) is not reasonable because it is established for the rested particle on the wall.

Example 2. 2　There is a spherical particle with $d_p = 20\mu m$ and $\rho_p = 2 \times 10^3 kg/m^3$. The gas density is $\rho = 1\text{kg/m}^3$. The duct diameter $d_D = 2R = 1\text{m}$.

(1) If the particle appears in gas stream, calculate the average gas velocity of keeping the particle suspension.

(2) If this particle is deposited on the smooth surface of a pipe, determine the maximum velocity in the core area of the pipe to ensure the particle on the surface to be blown up into the gas stream.

Solution

(1) According to equation (2.46), in order to keep the particle in suspension in gas stream, the shear lift force must be greater than the particle gravity, written as

$$L_s \geqslant m_p g$$

To ensure the particle to be suspended, the Reynolds number should be large enough. In this case, according to Fig. 2.5, the lift coefficient is

$$C_s = 0.15$$

Based on equations (2.45) and (2.46), we have

$$C_s \frac{\pi}{4} d_p^2 \frac{\rho v^2}{2} = \frac{\pi}{6} d_p^3 \rho_p g$$

Then, the average gas velocity of keeping the 20μm diameter particle suspension is

$$v = \left(\frac{8}{6} \frac{\rho_p}{C_s \rho} d_p g \right)^{1/2} = 3.41 \text{ (m/s)}$$

This value is over estimated because the Reynolds number is

$$\text{Re} = \frac{\rho d_p v}{\mu} = 3.7$$

It is shown that in order to keep 20μm particlesuspension, the gas velocity is $v<3.41\text{m/s}$.

(2) If this particle is deposited on the smooth surface of a pipe, according to equation (2.62), it is given by

$$L_p \geqslant m_p g$$

From equation (2.61), we have

$$\frac{1}{21} \pi \rho d_p^2 \left(\frac{d_p}{2R} \right)^{2/7} v_{max}^2 \geqslant \frac{\pi}{6} d_p^3 \rho_p g$$

To ensure the particle on the surface of the duct wall to be blown up into the gas stream, the maximum velocity in the core area of the pipe is

$$v_{max} = \sqrt{\frac{21}{6} \frac{\rho_p}{\rho} \left(\frac{d_p}{2R} \right)^{-2/7} d_p g} = 5.48 \text{ (m/s)}$$

It is clear that, from this example, to blow the particulate from the surface of the wall up to the gas is much more difficult than to keep the particulate to be suspended in gas because the aerodynamic force is much weaker near the wall.

It should be pointed out that in above calculation, for the large particle, the force of Van der Waals force was neglected. But for the fine particle, the Van der Waals force is remarkable.

This idea is important in design and application of aerodynamic transportation of particulate[7]. If the particulate is stationary on the wall, the blowing gas velocity is higher. While if the particulate as long as is blown up, the gas velocity should be reduced, as shown in Fig. 2. 7. In this case, it is not only less energy consumption, but also lower tube wall erosion.

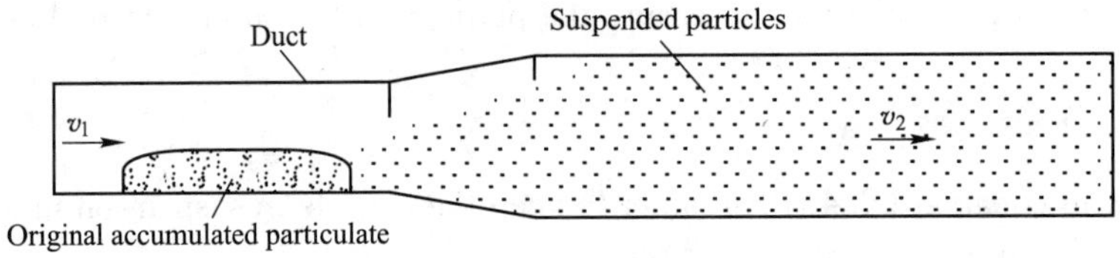

Fig. 2. 7 Velocity distribution in aerodynamic transportation of particulate in duct ($v_1 > v_2$)

Exercises

2. 1 Classify small particles of fly ash with density of 2×10^3kg/m^3. The particles are falling in a vertical tube against a rising current of air at standard conditions (air viscosity is $\mu = 1.85\times10^{-5}$Pa · s, and air density is $\rho = 1.2$kg/m^3). Calculate the minimum size of fly ash which will settle to the bottom of the tube if the air is rising through the tube at the velocity of 1m/s.

2. 2 What is the stop distance of a spherical particle of 10μm diameter and density 3×10^3kg/m^3 projected into still air at standard conditions with an initial velocity of 0. 1m/s?

2. 3 A spherical particle with 400μm in diameter is allowed to settle in air at standard conditions (air viscosity is $\mu = 1.85\times10^{-5}$Pa · s, and air density is $\rho = 1.2$kg/m^3). Calculate the particle' s terminal settling velocity. How far will the particle fall in 5min? The particle density is 3×10^3kg/m^3.

2. 4 A 100μm diameter spherical particle with initial velocity of 20m/s is ejected in to a gas flow in opposite direction where the gas velocity is 5m/s, as shown in Fig. 2. 8. If the gas viscosity is $\mu = 1.85\times10^{-5}$Pa · s, and the particle density is 2×10^3kg/m^3, calculate the moving distance of the particle in gas when the particle speed is reduced to zero.

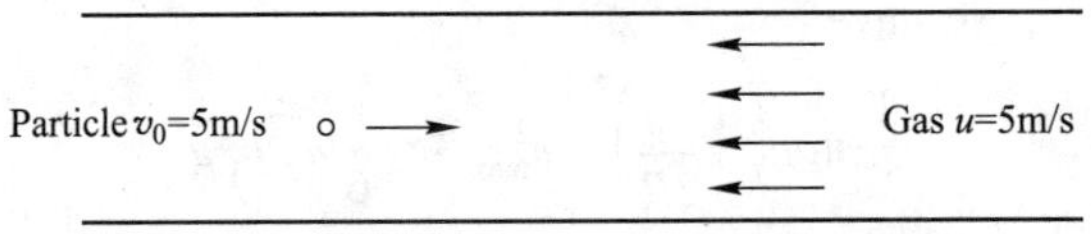

Fig. 2. 8 A particle motion in a gas flow

2. 5 A particle with 10μm in diameter falls suddenly in a vortex flow at $r = 0.1$m as shown in Fig. 2. 9. Suppose the gas velocity is $u = 20$m/s, which is uniform in the radius from r_1 to r_2. The gas density is $\rho = 1.2$kg/m^3 and the particle density is $\rho_p = 2\times10^3$kg/m^3. Calculate the time and turns of this particle touching the wall where $r_2 = 0.4$m.

2. 6 A 100μm diameter spherical particle has a density of 2×10^3 kg/m^3. The gas which has a density of 1kg/m^3 is flowing in a cylindrical tube with radius of 0. 5m. Determine the maximum gas velocity in the core area of the pipe to ensure this particle on the surface to be blow up.

2. 7 A 10μm diameter spherical particle in an alternating electric field with field strength of $E = 4\times10^5\sin(\pi/2)t$ has the charge of 10^{-16}C and density of 2×10^3kg/m^3.

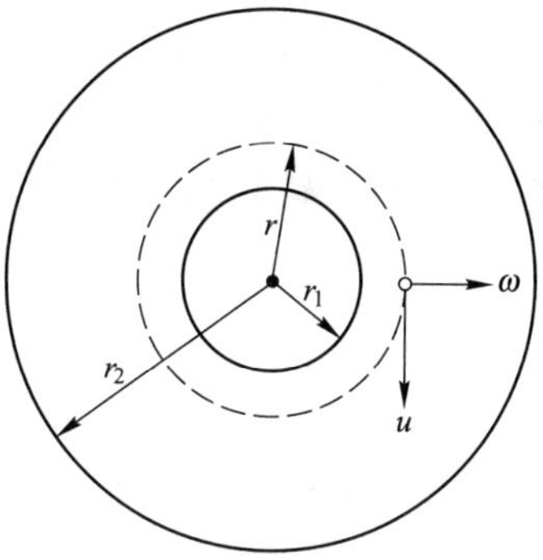

Fig. 2. 9 A particle motion in a vortex flow

(1) Calculate the maximum vibration velocity and swing at the standard condition.

(2) Plot the variation curve of the velocity in one period of the alternating electric field.

References

[1] Williams M M R, Loyalka S K. *Aerosol Science: Theory and Practice* [M]. New York, Wiley-Interscience, 2000.

[2] Happel J, Brenner H. *Low Reynolds Number Hydrodynamics: With Special Applications to Particulate* [M]. New York, Springer, 1983.

[3] Foias C, Manley O P, Rosa R, Temam R. *Navier-Stokes Equations and Turbulence* [M]. Cambridge University Press, 2001.

[4] White H J. *Industrial Electrostatic Precipitation* [M]. Massachusetts, Andison-Wesley, 1963.

[5] Mei R. An approximate expression for the shear lift force on a spherical particle at finite reynolds number [J]. International Journal of Multiphase Flow, 1992, 18 (1): 145-147.

[6] Martin C. *Air Pollution Control Theory* [M]. New York, Mcgraw-Hill, 1976.

[7] Mallick S S, Wypych P W. Minimum transport boundaries for pneumatic conveying of powders [J]. Powder Technology, 2009, 194 (3): 181-186.

3 Diffusion of Aerosol Particulate

The diffusion of fine particles is caused by the thermo-motion of fluid molecules. Particles suspended in a fluid undergo irregular random motion due to bombardment by surrounding fluid molecules. This is called Brownian motion. The result of this random motion always leads to the particles moving from the higher concentration to the lower concentration.

The problems discussed in particle diffusion are the transformation and concentration distribution of the fine particles (Particle size is less than 1μm). As we have known that, in particulate pollutant control, one of the most important problems is to find the particle concentration distribution in gas. In this way, we can evaluate the air cleaning performances of the pollution control devices.

The particle concentration distribution can be described by analytical solution or numerical solution. In the most cases, there are rare analytical solutions. We have to use the numerical solutions to describe the particle concentration distribution. Of course, if a concise concentration distribution formula can be developed, it will be very convenient in theoretical analysis and engineering applications.

3.1 Basic Diffusion Law

3.1.1 Fick's First Law

Basic mathematic model of describing the mass transformation process in a medium is the diffusion equation.

In homogeneous fluid, the mass flow rate per square meter is proportional to the grads of concentration. This is the well-known Fick's first law[1], which is given by

$$F = -D\frac{\partial c}{\partial x} \tag{3.1}$$

Where F——the flux in $kg/(m^2 \cdot s)$;

c——the mass concentration in kg/m^3;

D——the diffusion coefficient in m^2/s.

For spherical particles suspended in a perfect gas, D is estimated from the kinetic of gases, which is given by Stokes-Einstein as

$$D = k_B T C_c / 3\pi\mu d_p = kTB_p \tag{3.2}$$

Where k_B——Boltzmann constant, $k_B = 1.38 \times 10^{-23}$ J/K;

T——the absolute temperature in K;

μ——kinematic viscosity in Pa · s;

B——particle mobility in $m^2/(N \cdot s)$, written as

$$B_p = C_c/3\pi\mu d_p \tag{3.3}$$

Diffusion coefficient of particles with density of $1000kg/m^3$ at T=293K is given in Table 3.1.

Table 3.1 Diffusion coefficient and other parameters of particles with material density of $1000kg/m^3$ at T=293K

$d_p/\mu m$	C_c	$v/m \cdot s$	τ/s	$B_p/m^2 \cdot (N \cdot s)^{-1}$	$D/m^2 \cdot s^{-1}$
0.00037①	—	—	2.6×10^{-10}	9.7×10^{15}	1.8×10^{-5}
0.01	23.04	6.95×10^{-8}	7.1×10^{-9}	1.4×10^{13}	5.5×10^{-8}
0.1	2.866	8.65×10^{-7}	8.8×10^{-8}	1.7×10^{11}	6.8×10^{-10}
1	1.152	3.48×10^{-5}	3.5×10^{-6}	6.8×10^{9}	2.7×10^{-11}
10	1.015	3.06×10^{-3}	2.3×10^{-4}	6.0×10^{8}	2.4×10^{-12}
100	1.002	2.61×10^{-1}	1.3×10^{-2}	5.9×10^{7}	2.4×10^{-13}

① Means the diameter of molecules.

3.1.2 Fick's Second Law

When Fick's first law and the mass conservation law are applied, Fick's second law can be developed[2] by

$$\frac{\partial c}{\partial t} = D\nabla^2 c - \Delta(uc) \tag{3.4}$$

For three dimensional coordinates, if the gas flows with velocity u in x coordinate, the flowing term is much greater than the diffusion term. Equation (3.4) becomes

$$\frac{\partial c}{\partial t} = D\left(\frac{\partial^2 c}{\partial y^2} + \frac{\partial^2 c}{\partial z^2}\right) - \frac{\partial (uc)}{\partial x} \tag{3.5}$$

For cylindrical coordinates, the diffusion equation is expressed by

$$\frac{\partial c}{\partial t} = D\left(\frac{\partial^2 c}{\partial r^2} + \frac{1}{r}\frac{\partial c}{\partial r}\right) - \frac{\partial (uc)}{\partial x} \tag{3.6}$$

3.2 Diffusion in Still Gas

3.2.1 Diffusion for Reflection Wall in Still Gas

Suppose that there is a plane pollutant source at $x=0$ in a still gas, as shown in Fig. 3.1.

This is one dimensional problem. Because the gas is still, equation (3.4) is simplified by

$$\frac{\partial c}{\partial t} = D\frac{\partial^2 c}{\partial x^2} \tag{3.7}$$

The solution of this differential equation is

$$c = \frac{A}{\sqrt{t}} e^{-x^2/4Dt} \tag{3.8}$$

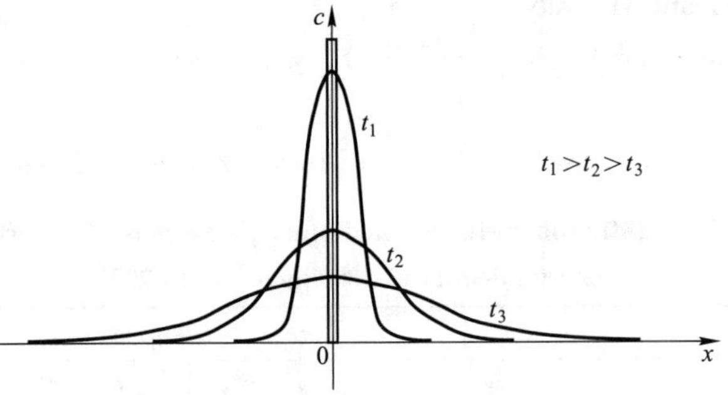

Fig. 3. 1 Concentration distribution of a plane source in infinite space

It is x axis of symmetry. To get a special solution, the constant A must be found, as $t>0$, $x\to\pm\infty$, and concentration $c\to 0$. Suppose the overall mass emitted from the plane source is given by

$$M = \int_{-\infty}^{+\infty} \frac{A}{\sqrt{t}} e^{-x^2/4Dt} dx = 2A\sqrt{D}\int_{-\infty}^{+\infty} e^{-x^2/4Dt} d\left(\frac{x}{\sqrt{4Dt}}\right) = 2A\sqrt{\pi D} \tag{3.9}$$

Then, equation (3. 8) becomes

$$c = \frac{M}{2\sqrt{\pi Dt}} e^{-x^2/4Dt} \tag{3.10}$$

If a wall without absorption effect is located at $x = 0$, the diffusion takes place in positive direction of x. In this case, the concentration distribution is expressed by

$$c = \frac{M}{\sqrt{\pi Dt}} e^{-x^2/4Dt} \tag{3.11}$$

Contrasting equation (3. 10) to equation (3. 11), for the plane source with mass emission of M, the concentration of one side diffusion is as twice as that of two side emission. In the other words, if there is only one side diffusion of a plane source, it looks like a ' reflecting wall' to be present and the concentration distribution is doubled. That is, the concentration follows the superposition principle.

Example 3. 1 There are two walls without absorption effect. The distance between these two walls is b. The pollutant mass of M is emitted only from the surface of the left side wall at $t = 0$, as shown in Fig. 3. 2. Find the concentration distribution between these two wall at any time t, and draw the curves of the distribution schematically.

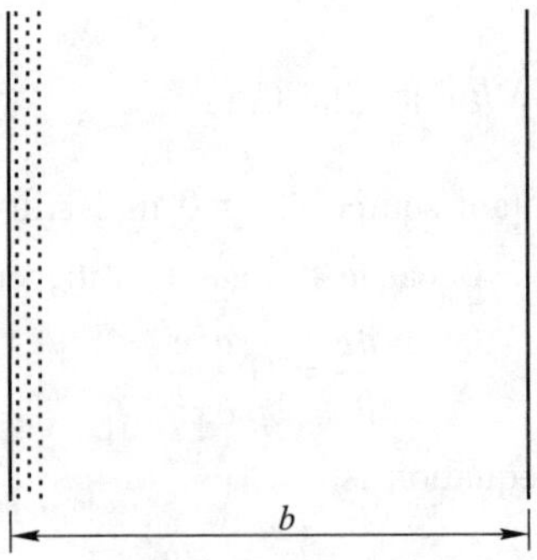

Fig. 3. 2 Pollutant emitted from the left infinite wall surface

Solution

First, move away the right wall. For the left side plane pollutant source, at time t, the concentration follows equation(3.11), written as

$$c_1 = \frac{M}{\sqrt{\pi Dt}} e^{-x^2/4Dt}$$

The curves of the distribution c_1 is indicated in Fig. 3.3.

Then, put back the right side wall. At time t, as long as the pollutant contacts the right side wall, the pollutant will be reflected, like a mirror. Behind the right wall, there would be an image plane source at $x = 2b$, as shown in Fig. 3.3. The dash dot line in the range of x from 0 to b is the concentration which needs to be added.

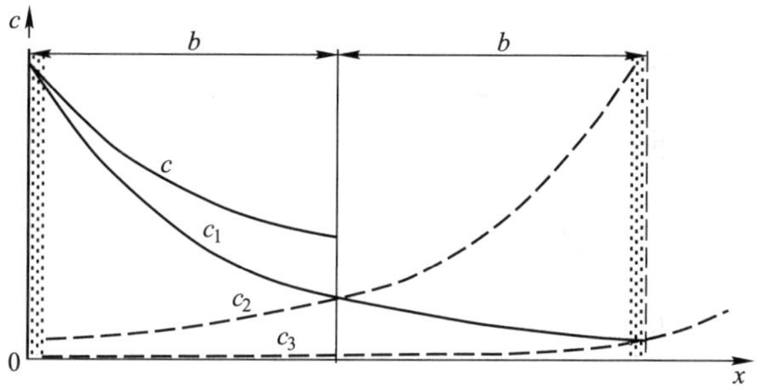

Fig. 3.3 Concentration distribution between these two wall at any time t

In order to determine the concentration contributed by the image plane source, the concentration distribution function (the dash dot line) in the range of x from 0 to $2b$ has to be found. Because the concentration distribution produced by the image plane pollutant source is symmetrical to the real one. Then write the concentration distribution function in the range from 0 to $2b$, written as

$$c_2 = \frac{M}{\sqrt{\pi Dt}} e^{-(2b-x)^2/4Dt}$$

According to the superposition principle, the concentration distribution between these two wall at any time t is

$$c = c_1 + c_2 = \frac{M}{\sqrt{\pi Dt}} e^{-x^2/4dt} + \frac{M}{\sqrt{\pi Dt}} e^{-(2b-x)^2/4Dt} \qquad 0 \leqslant x \leqslant b$$

The curves of the concentration distributions between these two wall at any time t are shown in Fig. 3.3.

Readers may envision that, when the concentration distribution caused by the real source contacts the image source wall at $x = 2b$, the reflection will take place again. It just like there is another image source at $x = 4b$. This need to be added again, called second superposition. The concentration distribution is

$$c_3 = \frac{M}{\sqrt{\pi Dt}} e^{-(4b-x)^2/4Dt}$$

Then, in the range of $0 \sim b$, The concentration distribution is written as

$$c = c_1 + c_2 + c_3 = \frac{M}{\sqrt{\pi D t}} e^{-x^2/4dt} + \frac{M}{\sqrt{\pi Dt}} e^{-(2b-x)^2/4Dt} + \frac{M}{\sqrt{\pi Dt}} e^{-(4b-x)^2/4Dt} \quad 0 \leqslant x \leqslant b$$

By parity of reasoning, n times of additions are needed, till $n \rightarrow \infty$. In this way, the concentration distribution between these two wall may tend towards the precise result. In fact, just need one time of superposition to get quite precise concentration distribution between these two wall.

3.2.2 Diffusion for Absorption Wall in Still Gas

There is a vertical absorption wall. At $x = 0$, the right side of this wall is adjacent to an infinite medium with pollutant materials, as shown in Fig. 3.4. When $t = 0$, the initial concentration is uniform, $c = c_0$.

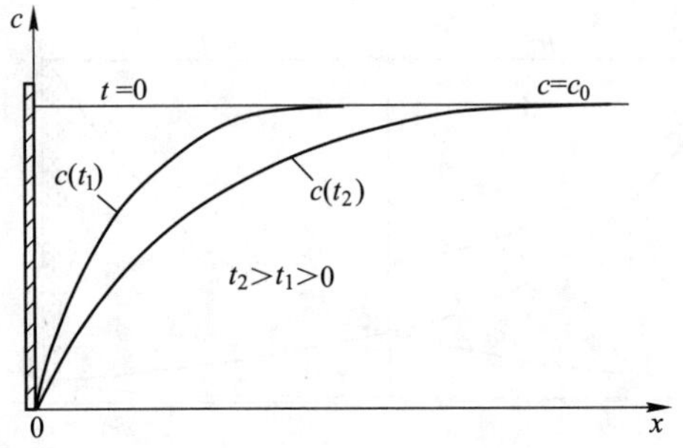

Fig. 3.4 Concentration distribution of the pollutant near the absorption wall

This is still one dimensional problem. That is, the pollutant diffusion follows equation (3.7). For absorption wall, the characteristics is that, on the wall surface, the concentration is zero. Therefore, the boundary condition is

$$t = 0,\ x > 0,\quad c = c_0;\qquad x = 0,\ t > 0,\quad c = 0 \tag{3.12}$$

This problem has precise solution, written as

$$c = \frac{2c_0}{\sqrt{4\pi Dt}} \int_0^x e^{-\xi^2} d\xi = \frac{2c_0}{\sqrt{\pi}} \int_0^{x/\sqrt{4Dt}} e^{-\eta^2} d\eta = c_0 \operatorname{erf}\left(\frac{x}{\sqrt{4Dt}}\right) \tag{3.13}$$

where, $\eta = x/\sqrt{4Dt}$.

In equation (3.13), erf is called error function.

We are more interested in the particulate mass deposited on the per square meter of the wall. Fick' s first law is given by

$$F = -D\left(\frac{\partial c}{\partial x}\right)_{x=0} = -D\frac{\partial}{\partial x}\left[\frac{2c_0}{\sqrt{4\pi Dt}} \int_0^x e^{-\xi^2} d\xi\right] = -c_0\sqrt{D/\pi t} \tag{3.14}$$

The negative in equation (3.14) indicates that the direction of diffusion deposition is adverse to the direction of x axes.

The particulate mass deposited on per square meter of the absorption wall at anyperiod of time $t_0 \sim t$ can also be caculated by

$$\dot{m} = -\int_{t_0}^{t} F\mathrm{d}t = 2c_0\sqrt{\frac{D(t - t_0)}{\pi}} \tag{3.15}$$

Where $\dot{m}$ ——the particulate mass deposited on per square meter of the absorption wall.

In practice, particulate pollutant or gaseous pollutant are commonly not absorbed in an infinite space. Of course, there is no infinite absorption wall either. However, the diffusion equation of the absorption wall in the infinite space can be used to solve the diffusion problem in a limited space.

Example 3.2 There is a rectangular container with a in width, b in thickness, and h in height respectively, as shown in Fig. 3.5. Suppose the left and right sides are absorption walls, and the others are not. The initial pollutant concentration is c_0. Discuss the concentration distribution variation with time in this container.

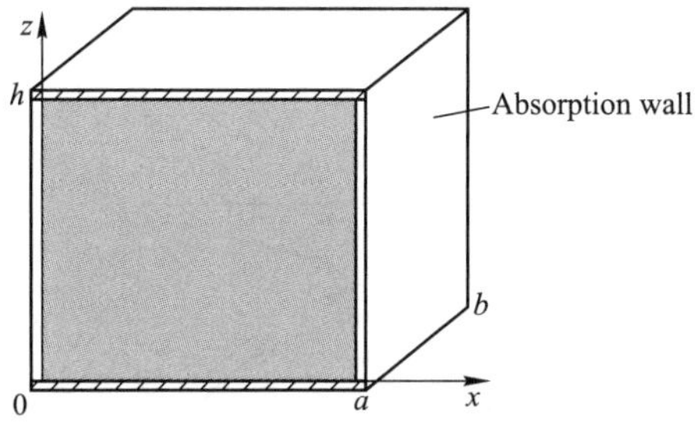

Fig. 3.5 Pollutant diffusion in a rectangular container with two absorption walls

Solution

Because the four plates in front, back, top and bottom of the container are not absorption walls, the concentration is not the function of y and z. This is one dimensional problem. The diffusion equation is equation (3.7). The respected boundary condition is

$$t = 0,\ x > 0,\ c = c_0;\quad x = 0,\ t > 0,\ c = 0;\quad x = a,\ t > 0,\ c = 0 \tag{3.16}$$

This question has the series solution, written as

$$c = \frac{2c_0}{n\pi}\sum_{n=1}^{\infty}[1 - (-1)^n]\sin\left(\frac{n\pi x}{a}\right)\exp\left[-\left(\frac{n\pi}{a}\right)^2 Dt\right] \tag{3.17}$$

Equation (3.17) is a precise solution of this problem. However, it is too cumbrous to be calculated. To simplify this problem, use the concentration distribution equation (3.13) to analyze the concentration distribution in a container with absorption walls.

If we take away the right side absorption wall first, at time t, the absorption effect of the left side follows the error function distribution equation (3.13). It is written as

$$c_1 = c_0 \operatorname{erf}\left(\frac{x}{\sqrt{4Dt}}\right) \tag{3.18}$$

The concentration distribution c_1 is illustrated in Fig. 3.6.

If we keep the right side wall and move away the left side wall, the concentration distribution c_2 looks like the curve 2 in Fig. 3.6. The curve 2 and curve 1 are symmetrical. The concentration of any point x on curve 2 is equal to the concentration of the point (a-x) on curve 1. Then the concentration distribution of curve 2 is given by

$$c_2 = c_0 \operatorname{erf}\left(\frac{a - x}{\sqrt{4Dt}}\right) \tag{3.19}$$

The reduced value of the concentration caused by the right side at x is written as

$$\Delta c = c_0 \operatorname{erf}\left(\frac{a}{\sqrt{4Dt}}\right) - c_2 = c_0\left[\operatorname{erf}\left(\frac{a}{\sqrt{4Dt}}\right) - \operatorname{erf}\left(\frac{a-x}{\sqrt{4Dt}}\right)\right] \tag{3.20}$$

According to the superposition principle, the concentration distribution due to the two absorption walls is

$$c = c_1 - \Delta c = c_0\left[\operatorname{erf}\left(\frac{x}{\sqrt{4Dt}}\right) + \operatorname{erf}\left(\frac{a-x}{\sqrt{4Dt}}\right) - \operatorname{erf}\left(\frac{a}{\sqrt{4Dt}}\right)\right] \tag{3.21}$$

The concentration distribution curve 3 of the co-action of two absorption walls in a rectangular container is shown in Fig. 3. 6. It is easy to prove that equation (3. 21) satisfy the boundary condition (3. 16). Obviously, equation (3. 21) is much simpler and more applicable than equation (3. 17).

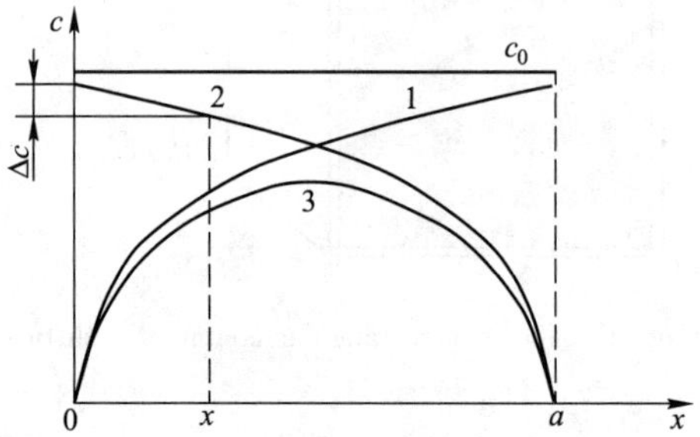

Fig. 3. 6 Concentration distributions in a rectangular container with two absorption walls

In the case of four absorption walls, or even six absorption walls, the solution of the diffusion equation is more complicated. But by means of above method, the problem becomes very simple. For example, for the four absorption walls in a rectangular container, such as the left, right, front, and back wall, if the y axes is perpendicular to the paper, it is a problem of two dimensional diffusion. The concentration distribution direction can be written directly as

$$c = c_0\left[\operatorname{erf}\left(\frac{x}{\sqrt{4Dt}}\right) + \operatorname{erf}\left(\frac{a-x}{\sqrt{4Dt}}\right) - \operatorname{erf}\left(\frac{a}{\sqrt{4Dt}}\right)\right]\left[\operatorname{erf}\left(\frac{y}{\sqrt{4Dt}}\right) + \operatorname{erf}\left(\frac{b-y}{\sqrt{4Dt}}\right) - \operatorname{erf}\left(\frac{b}{\sqrt{4Dt}}\right)\right] \tag{3.22}$$

3. 2. 3 Diffusion for Absorption Surface of a Sphere in Still Gas

In an infinite still medium, under the spherical coordinates, the diffusion equation of pollutant diffusion toward the surface of sphere is

$$\frac{\partial c}{\partial t} = D\left(\frac{\partial^2 c}{\partial r^2} + \frac{2}{r}\frac{\partial c}{\partial r}\right) \tag{3.23}$$

The boundary conditions are

$$t = 0,\ r > R,\ c = c_0;\quad r = R,\ t > 0,\ c = 0 \tag{3.24}$$

Where R——the radius of the sphere;

c_0——the initial pollutant concentration in gas.

This problem has a general solution which is given by

$$c = A\left[1 - \frac{1}{2^2}(\lambda r)^2 + \frac{1}{2^2 4^2}(\lambda r)^4 - \cdots\right] r^{-1/2}\exp(-\lambda^2 Dt) \tag{3.25}$$

This series solution is convergent as $r \to \infty$[3]. However, it is very difficult to determine the constants of A and λ according to the boundary conditions given by equation (3.24). Now, we use the variable substitution to get the analytical solution which is given as follows.

Equation (3.13) reminds that $r/\sqrt{4Dt}$ could most possibly be the function of the solution of equation (3.23). Therefore, we introduce a variable of $\eta = r/\sqrt{4Dt}$, that is, $c = c(\eta)$, then

$$\frac{\partial c}{\partial t} = \frac{\partial c}{\partial \eta}\frac{\partial \eta}{\partial t} = -\frac{\eta}{2t}\frac{\partial c}{\partial \eta} \tag{3.26}$$

$$\frac{\partial c}{\partial r} = \frac{1}{\sqrt{4Dt}}\frac{\partial c}{\partial \eta} \tag{3.27}$$

$$\frac{\partial^2 c}{\partial r^2} = \frac{1}{4Dt}\frac{\partial^2 c}{\partial \eta^2} \tag{3.28}$$

Substituting above three equations into equation (3.23), after rearranging, we obtain

$$\frac{\partial^2 c}{\partial \eta^2} + \left(\frac{2}{\eta} + 2\eta\right)\frac{\partial c}{\partial \eta} = 0 \tag{3.29}$$

Let

$$p = \frac{\partial c}{\partial \eta},\ \frac{\partial p}{\partial \eta} = \frac{\partial^2 c}{\partial \eta^2}$$

Equation (3.29) becomes

$$\frac{\partial p}{\partial \eta} + \left(\frac{2}{\eta} + 2\eta\right)p = 0 \tag{3.30}$$

The solution of equation (3.30) is

$$p = A\exp(-2\ln\eta - \eta^2)$$

Thus

$$\frac{\partial c}{\partial \eta} = A\frac{1}{\eta^2}\exp(-\eta^2) \tag{3.31}$$

The solution of equation (3.31) is

$$c = A\left[-\frac{1}{\eta}\exp(-\eta^2) - 2\int_0^\eta \exp(-\eta^2)\,d\eta\right] + B \tag{3.32}$$

We noticed that

$$2\int_0^\eta \exp(-\eta^2)\,d\eta = \sqrt{\pi}\,\mathrm{erf}(\eta) \tag{3.33}$$

Then, equation (3.32) is simplified as

$$c = A\left[-\frac{1}{\eta}\exp(-\eta^2) - \sqrt{\pi}\,\mathrm{erf}(\eta)\right] + B \tag{3.34}$$

Using boundary condition, at $r = R$, $\eta = R/\sqrt{4Dt}$, and $c = 0$, from equation (3.34), we have

$$A\left[-\frac{\sqrt{4Dt}}{R}\exp\left(-\frac{R^2}{4Dt}\right) - \sqrt{\pi}\,\mathrm{erf}\left(\frac{R}{\sqrt{4Dt}}\right)\right] + B = 0 \tag{3.35}$$

As $r \to \infty$, then $c = c_0$, $\frac{1}{\eta}\exp(-\eta^2) = 0$, and $\mathrm{erf}(\eta) = 1$, equation (3.34) becomes

$$-\sqrt{\pi}A + B = c_0 \tag{3.36}$$

Therefore, the constants of A and B can be determined. The pollutant concentration distribution in still gas caused by the absorption surface of sphere is

$$c = c_0 \frac{\sqrt{4Dt/\pi}\left[\frac{1}{r}\exp\left(-\frac{r^2}{4Dt}\right) - \frac{1}{R}\exp\left(-\frac{R^2}{4Dt}\right)\right] + \mathrm{erf}\left(\frac{r}{\sqrt{4Dt}}\right) - \mathrm{erf}\left(\frac{R}{\sqrt{4Dt}}\right)}{1 - \mathrm{erf}\left(\frac{R}{\sqrt{4Dt}}\right) - \frac{\sqrt{4Dt/\pi}}{R}\exp\left(-\frac{R^2}{4Dt}\right)} \quad r \geqslant R \tag{3.37}$$

Equation (3.37) is a precise solution of equation (3.23) in the rage of $[R, \infty]$. Since D is very small and $r \geqslant R$, equation (3.37) can be further simplified approximately as

$$c = c_0\left[\mathrm{erf}\left(\frac{r}{\sqrt{4Dt}}\right) - \mathrm{erf}\left(\frac{R}{\sqrt{4Dt}}\right)\right] \Big/ \left[1 - \mathrm{erf}\left(\frac{R}{\sqrt{4Dt}}\right)\right] \tag{3.38}$$

If R is small, equation (3.38) is simplified as

$$c = c_0\left[\mathrm{erf}\left(\frac{r}{\sqrt{4Dt}}\right) - \mathrm{erf}\left(\frac{R}{\sqrt{4Dt}}\right)\right] = c_0 \mathrm{erf}\left(\frac{r-R}{\sqrt{4Dt}}\right) \quad r \geqslant R \tag{3.39}$$

Circumspective reader may notice that equation (3.39) is the precise solution of $\frac{\partial c}{\partial t} = D\frac{\partial^2 c}{\partial r^2}$ because in equation (3.23), $\frac{\partial^2 c}{\partial r^2} \gg \frac{2}{r}\frac{\partial c}{\partial r}$.

The concentration distribution of the pollutant diffusion to the surface of a sphere is illustrated in Fig. 3.7. The pollutant diffusion to the surface of a sphere is common in particulate pollutant control, such as granular bed filter, water sprayer, and ball packed tower.

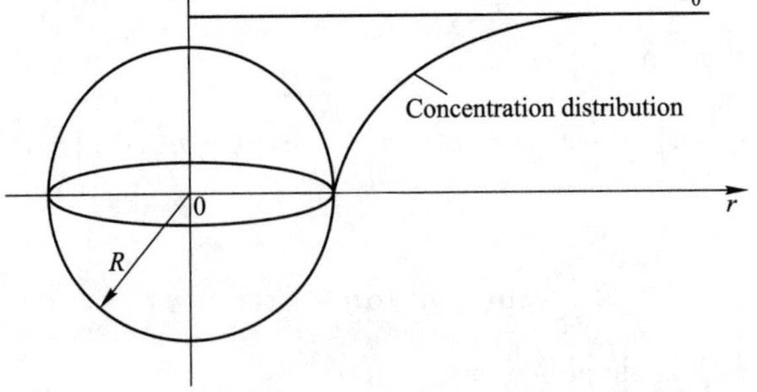

Fig. 3.7 Concentration distribution of the diffusion for absorption surface of a sphere in still gas

3.2.4 Diffusion for Absorption Surface of a Cylindrical Tube in Still Gas

There are two cases of the diffusion toward a cylindrical tube. One is the diffusion toward the outside surface of the tube. The other is the diffusion toward the inside surface of the tube.

In cylindrical coordinates, in still gas, equation (3.6) becomes

$$\frac{\partial c}{\partial t} = D\left(\frac{\partial^2 c}{\partial r^2} + \frac{1}{r}\frac{\partial c}{\partial r}\right) \tag{3.40}$$

The boundary conditions are

$$t = 0, \ r > R, \ c = c_0; \quad r = R, \ t > 0, \ c = 0 \tag{3.41}$$

Because of $\frac{\partial^2 c}{\partial r^2} >> \frac{1}{r}\frac{\partial c}{\partial r}$, equation (3.40) can be written as

$$\frac{\partial c}{\partial t} = D\left(\frac{\partial^2 c}{\partial r^2}\right) \tag{3.42}$$

The solution of the concentration distribution due to the diffusion toward the outside surface of the tube can be derived from equation (3.42) and boundary condition (3.41) as

$$c = c_0 \mathrm{erf}\left(\frac{r - R}{\sqrt{4Dt}}\right) \qquad r \geqslant R \tag{3.43}$$

Similarly, the concentration distribution due to the diffusion toward the inside surface of the tube is

$$c = c_0 \mathrm{erf}\left(\frac{R - r}{\sqrt{4Dt}}\right) \qquad 0 < r \leqslant R \tag{3.44}$$

From Section 3.2.2 to Section 3.2.4, we have observed that in normal direction of the absorption surface in an infinite space, the concentration at any x point on the line which is perpendicular to the absorption surface can be expressed as

$$c \approx c_0 \mathrm{erf}\left(\frac{x}{\sqrt{4Dt}}\right) \tag{3.45}$$

Where x——the distance from the any point in the space to the absorption surface.

Thus, it can be expanded that, generally, the concentration distribution near any irregular surface approximately follows equation (3.45). This conclusion is very useful in pollutant absorption calculation.

Example 3.3 There is a packed tower as shown in Fig. 3.8. Fifty pieces of the corrugated plates are used for the packed material. The effective area of each plate is 2m^2. Suppose the pollutant concentration around the plates keeps no change, that is $c = c_0$. Calculate the mass of the pollutant collected on the plates after time t.

Solution

The concentration near the plate is

$$c \approx c_0 \mathrm{erf}\left(\frac{x}{\sqrt{4Dt}}\right)$$

According to Fick's first law, the pollutant mass deposited on the per square meter and per time is

$$F = -D\left(\frac{\partial c}{\partial x}\right)_{x=0} = -c_0\sqrt{D/\pi t}$$

The pollutant mass deposited on the per square meter of the absorption wall from time 0 to t is

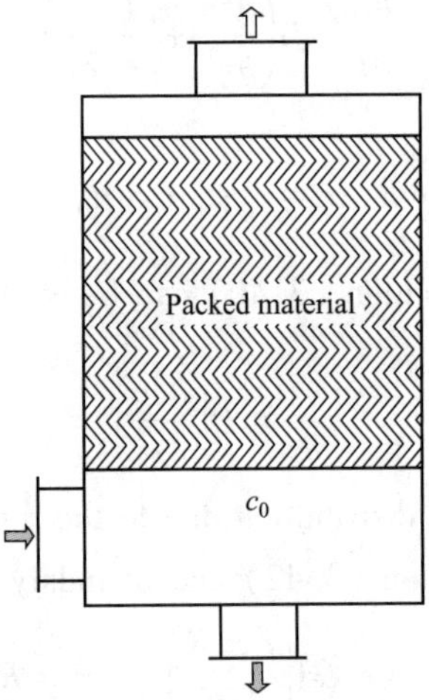

Fig. 3. 8　Schematic packed tower

$$\dot{m} = -\int_0^t F\mathrm{d}t = \int_0^t c_0\sqrt{D/\pi t}\,\mathrm{d}t = 2c_0\sqrt{\frac{Dt}{\pi}}$$

The total area of the corrugated plates is 100m^2. Then, the total mass of the pollutant collected on the corrugated plates after time t is

$$m = 200c_0\sqrt{\frac{Dt}{\pi}}$$

3. 3　Diffusion in Static Gas Flow

In a stable industrial production process, the gas flow parameters do not change considerably with time. Therefore, the gas flow state is thought to be a static flow.

3. 3. 1　Diffusion for Reflection Wall of a Rectangular Duct

In the steady state flow, in equation (3. 5), $\partial c/\partial t = 0$. Suppose the velocity u is uniform, equation (3. 5) becomes

$$u\frac{\partial c}{\partial x} = D\left(\frac{\partial^2 c}{\partial y^2} + \frac{\partial^2 c}{\partial z^2}\right) \tag{3.46}$$

This equation's solution is given by Gaussian formula

$$c = Kx^{-1}\exp\left[-\frac{u}{4Dx}(y^2 + z^2)\right] \tag{3.47}$$

Now, the constant K in equation (3. 47) must determined first to get special solution. For the diffusion of a point source in an infinite space, if the pollutant mass emitted per time is M, we have

$$M = \int_{-\infty}^{\infty}\int_{-\infty}^{\infty} uc\mathrm{d}y\mathrm{d}z = \int_{-\infty}^{\infty}\int_{-\infty}^{\infty} uKx^{-1}\exp\left[-\frac{u}{4Dx}(y^2 + z^2)\right]\mathrm{d}y\mathrm{d}z \tag{3.48}$$

Let

$$r^2 = y^2 + z^2$$

Eequation (3.48) becomes

$$M = uKx^{-1}\int_0^{2\pi}\int_0^{\infty}\exp\left(-\frac{u}{4Dx}r^2\right)\mathrm{d}r\mathrm{d}\theta = 4DK\pi \tag{3.49}$$

Then, the concentration distribution of a point source in an infinite space due to the diffusion of the pollutant is

$$c = \frac{M}{4\pi Dx}\exp\left[-\frac{u}{4Dx}(y^2 + z^2)\right] \tag{3.50}$$

Equation (3.50) is also the diffusion model of the smoke emitted from a stack in the atmosphere.

Now we discuss the concentration distribution of a point source effected by the reflection walls. Fig. 3.9 shows that a point source is located between two infinite reflection walls.

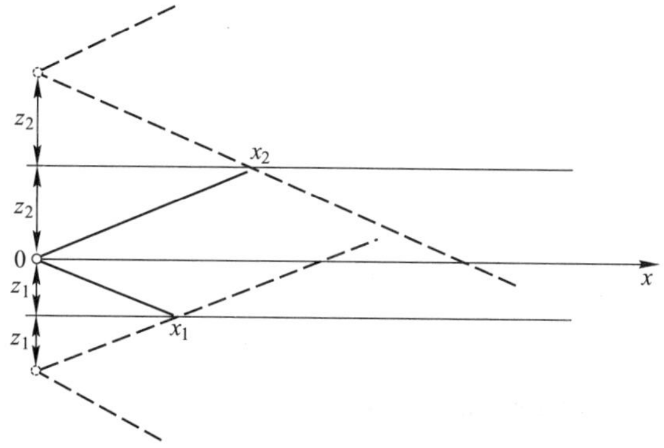

Fig. 3.9 Diffusion of a point source between two infinite reflection plates

If there is no wall, the concentration distribution in the space produced by this point source is described by equation (3.50). If the walls present, the original diffusion pollutants are going to contact with the walls. Then the pollutants will be reflected. This reflection effect will increase the concentration between the walls. In order to determine these concentration increments, two image point sources are assumed to be located outside of two walls. The concentrations caused by the real source and the two image sources without the wall are written as

$$c_0 = \frac{M}{4\pi Dx}\exp\left[-\frac{u}{4Dx}(y^2 + z^2)\right] \tag{3.51}$$

$$c_1 = \frac{M}{4\pi Dx}\exp\left\{-\frac{u}{4\pi Dx}[y^2 + (z + 2z_1)^2]\right\} \tag{3.52}$$

$$c_2 = \frac{M}{4\pi Dx}\exp\left\{-\frac{u}{4\pi Dx}[y^2 + (z - 2z_2)^2]\right\} \tag{3.53}$$

According to the superposition principle, the concentration distribution between these two reflection walls is

$$c = c_0 + c_1 + c_2 \tag{3.54}$$

It is clear that more times of superposition are needed as the x increase. In fact, a quite precise approximate concentration value will be obtained when just one time of superposition is used.

If the polluted gas flows inside arectangular duct, there are four reflection walls. Another two image point sources are added on the y coordinate at y_1 and y_2, expressed by

$$c_3 = \frac{M}{4\pi Dx}\exp\left\{-\frac{u}{4\pi Dx}[(y + 2y_1)^2 + z^2]\right\} \tag{3.55}$$

$$c_4 = \frac{M}{4\pi Dx}\exp\left\{-\frac{u}{4\pi Dx}[(y - 2y_2)^2 + z^2]\right\} \tag{3.56}$$

Therefore, the concentration distribution in a rectangular duct is

$$c = c_0 + c_1 + c_2 + c_3 + c_4 \tag{3.57}$$

3.3.2 Diffusion for Reflection Wall of a Cylindrical Duct

For a cylindrical tube with radius R, if a point source is located in the axes of the cylindrical tube as shown in Fig. 3.10 (If a point source is not located in the axes of the cylindrical tube, the problem becomes complex).

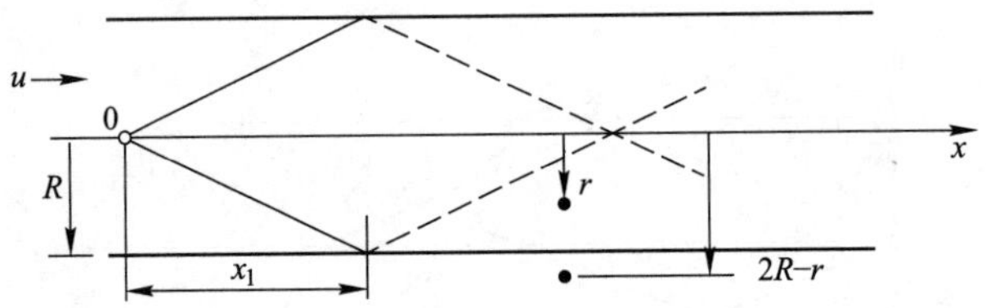

Fig. 3.10 Diffusion of a point source at the axial line of a cylindrical tube

Suppose the velocity u is uniform, the concentration distribution caused by a point source in an infinite space is

$$c_0 = \frac{M}{4\pi Dx}\exp\left(-\frac{u}{4Dx}r^2\right) \tag{3.58}$$

Making use of the symmetry, the contribution of the image source to the concentration of the point r in reflect area is equal to the concentration value of the point $2R-r$ produced by the real source. That is

$$c_1 = \frac{M}{4\pi Dx}\exp\left[-\frac{u}{4Dx}(2R - r)^2\right] \tag{3.59}$$

Then, the pollutant concentration distribution in a cylindrical tube due to the diffusion in the steady state gas flow is given by

$$c = c_0 + c_1 \tag{3.60}$$

3.3.3 Diffusion for Absorption Wall of a Rectangular Duct

Assume that the velocity in a rectangular duct with width a and height h is uniform and steady, as shown in Fig. 3.11. If the left side wall $y = 0$ and the right side wall $y = a$ have absorption effect, in this case, the diffusion equation (3.5) becomes

$$u\frac{\partial c}{\partial x} = D\left(\frac{\partial^2 c}{\partial y^2}\right) \tag{3.61}$$

If the duct is long enough, as the boundary condition, the concentration c_0 on the inlet section

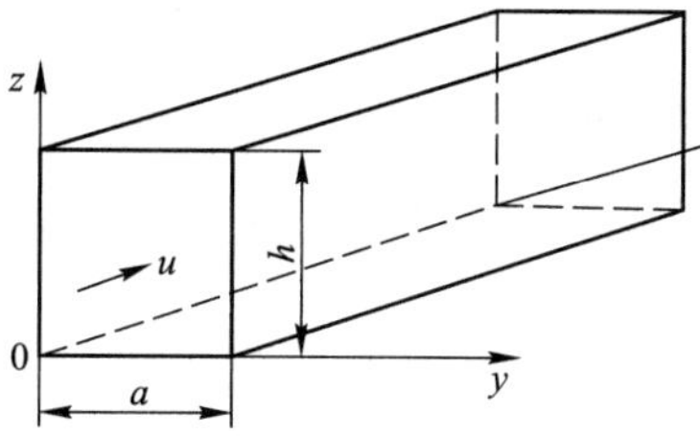

Fig. 3.11 Pollutant diffusion in a rectangular duct at steady state of gas flow

is assumed to be uniform, and the concentrations on the left and the right walls are zero due to the absorption wall, which is

$$x = 0,\ c = c_0;\quad y = 0,\ c = 0;\quad y = a,\ c = 0 \tag{3.62}$$

Let

$$\tau = x/u \tag{3.63}$$

Equation (3.61) becomes

$$\frac{\partial c}{\partial \tau} = D\frac{\partial^2 c}{\partial y^2} \tag{3.64}$$

The boundary condition becomes

$$\tau = 0,\ c = c_0;\quad y = 0,\ c = 0;\quad y = a,\ c = 0 \tag{3.65}$$

It is noticed that equation (3.64) and its boundary condition (3.65) are the same as equation (3.7) and boundary condition (3.16). The solution of (3.64) must be the same as equation (3.21), written as

$$c = c_0\left[\mathrm{erf}\left(\frac{y}{\sqrt{4D\tau}}\right) + \mathrm{erf}\left(\frac{a-y}{\sqrt{4D\tau}}\right) - \mathrm{erf}\left(\frac{a}{\sqrt{4D\tau}}\right)\right] \tag{3.66}$$

If four walls of the duct are allabsorption walls, the solution must be similar to equation (3.22), wnitten as

$$c = c_0\left[\mathrm{erf}\left(\frac{y}{\sqrt{4D\tau}}\right) + \mathrm{erf}\left(\frac{a-y}{\sqrt{4D\tau}}\right) - \mathrm{erf}\left(\frac{a}{\sqrt{4D\tau}}\right)\right]\left[\mathrm{erf}\left(\frac{z}{\sqrt{4D\tau}}\right) + \mathrm{erf}\left(\frac{h-z}{\sqrt{4D\tau}}\right) - \mathrm{erf}\left(\frac{h}{\sqrt{4D\tau}}\right)\right] \tag{3.67}$$

where, τ is decided by equation (3.63).

3.3.4 Diffusion for Absorption Wall of a Cylindrical Duct

If the polluted gas flows at uniform velocity of u constantly in a cylindrical tube with radius of a, and the inside surface of the cylindrical tube has the absorption effect, as shown in Fig. 3.12, the diffusion equation is expressed by

$$u\frac{\partial c}{\partial x} = D\left(\frac{\partial^2 c}{\partial r^2} + \frac{1}{r}\frac{\partial c}{\partial r}\right) \tag{3.68}$$

The boundary condition is

$$x = 0,\ c = c_0;\quad r = a,\ c = 0 \tag{3.69}$$

If let

$$\tau = x/u \tag{3.70}$$

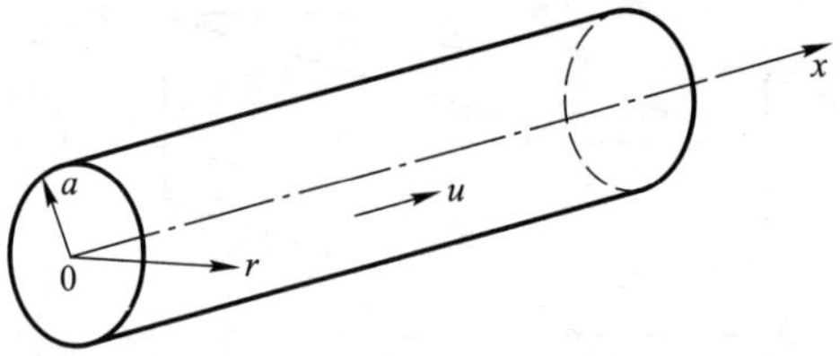

Fig. 3. 12　Pollutant diffusion in a cylindrical tube at steady state of gas flow

Equation (3. 68) and its boundary condition (3. 69) become

$$\frac{\partial c}{\partial \tau} = D\left(\frac{\partial^2 c}{\partial r^2} + \frac{1}{r}\frac{\partial c}{\partial r}\right) \tag{3.71}$$

$$\tau = 0, \ c = c_0; \ r = a, \ c = 0 \tag{3.72}$$

Because equation (3. 71) and its boundany condition (3. 72) are the same as equation (3. 40) and boudary condition (3. 41), the solution of (3. 71) must be the same expression as equation (3. 44), written as

$$c = c_0 \mathrm{erf}\left(\frac{R - r}{\sqrt{4D\tau}}\right) \quad 0 < r \leqslant R \tag{3.73}$$

It could be possible that the outside surface of the cylindrical tube has the absorption effect either, the concentration distribution due to the diffusion toward the outside surface of the cylindrical tube can be written as

$$c = c_0 \mathrm{erf}\left(\frac{r - R}{\sqrt{4D\tau}}\right) \quad r \geqslant R \tag{3.74}$$

3. 4　Diffusion of a Gas Flow Around an Axisymmetric Body

To describe the pollutant diffusion on the flow about a body is sometime quite difficult because the velocity around the body is variable. Here, we just discuss the diffusion on the flow around a cylinder and a sphere in a uniform gas stream.

3. 4. 1　Diffusion of a Gas Flow Around a Cylinder

Assume that the direction of a flow is perpendicular to the fine long cylinder with radius of a at low velocity, as shown in Fig. 3. 13.

The velocity distribution around the fine cylinder is determined by the Reynolds number, giuen by

$$\mathrm{Re} = \frac{\rho v_0 d_\mathrm{f}}{\mu} \tag{3.75}$$

Where　v_0——the gas velocity far from the isolated cylinder;

d_f——the diameter of the cylinder, $d_f = 2a$.

If Re $\leqslant 1$, that is, v_0 is very low, it is thought to be viscous flow, or Stokes' flow, around the cylinder. The stream function in viscous flow in polar coordinates is expressed as

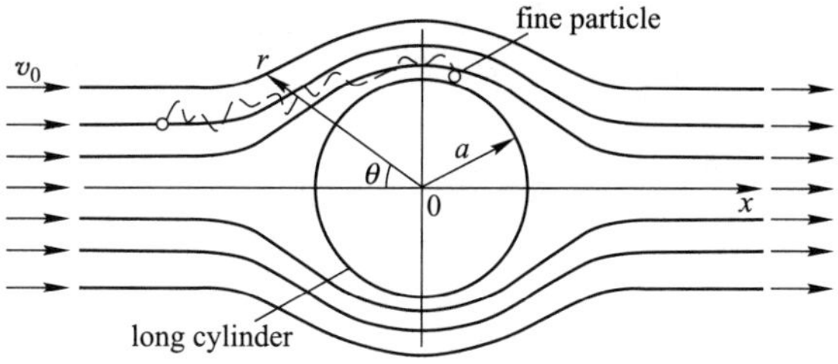

Fig. 3.13 Gas flow about a cylinder

$$\psi = \frac{v_0}{2\mathrm{La}}\left[2\ln\left(\frac{r}{a}\right) - 1 + \left(\frac{a}{r}\right)^2\right] r\sin\theta \tag{3.76}$$

Where La——Lamb constant, which is given by

$$\mathrm{La} = 2 - \ln\mathrm{Re} \tag{3.77}$$

The radial and tangential velocity components are given in terms of the stream function

$$v_r = \frac{\partial\psi}{r\partial\theta}, \quad v_\theta = -\frac{\partial\psi}{\partial r} \tag{3.78}$$

Using equations (3.76) and (3.78), the radial and tangential velocity components become

$$v_r = \frac{v_0}{2\mathrm{La}}\left[2\ln\left(\frac{r}{a}\right) - 1 + \left(\frac{a}{r}\right)^2\right]\cos\theta \tag{3.79}$$

$$v_\theta = -\frac{v_0}{2\mathrm{La}}\left[2\ln\left(\frac{r}{a}\right) + 1 - \left(\frac{a}{r}\right)^2\right]\sin\theta \tag{3.80}$$

For particulate control, we are just interested in the diffusionin normal (radial) direction of the 'absorption wall' of the long cylinder. Therefore, the diffusion in tangential direction is neglected. Another reason of no consideration of the diffusion in tangential direction is that the mass transportation of the tangential velocity is much stronger than that of the diffusion in tangential direction. Thus, the diffusion equation in polar coordinates is simplified as

$$\frac{\partial c}{\partial t} = D\left(\frac{\partial^2 c}{\partial r^2} + \frac{1}{r}\frac{\partial c}{\partial r}\right) - \frac{\partial(-v_r c)}{\partial r} \tag{3.81}$$

In equation (3.81), $-v_r$ means that the direction of the radial velocity opposites to the radial direction. In steady state flow, $\frac{\partial c}{\partial t} = 0$, and $\frac{\partial^2 c}{\partial r^2} >> \frac{1}{r}\frac{\partial c}{\partial r}$, equation (3.81) becomes

$$D\frac{\partial^2 c}{\partial r^2} + \frac{\partial(v_r c)}{\partial r} = 0 \tag{3.82}$$

It is still difficult to find the solution of equation (3.82) because the formula (3.79) of v_r is comparatively complicated. In order to find the solution of equation (3.82), the 'average value' of v_r is used in equation (3.79). Then, equation (3.82) can be written as

$$r\frac{\partial^2 c}{\partial r^2}+\frac{\bar{v}_r}{D}\frac{\partial c}{\partial r}=0 \tag{3.83}$$

The boundary condition is

$$r=a,\ c=0;\ r\to\infty,\ c=c_0 \tag{3.84}$$

The solution of (3.83) is

$$c=c_0\left\{1-\exp\left[-\frac{\bar{v}_r}{D}(r-a)\right]\right\} \tag{3.85}$$

Now we need to determine the average velocity $\bar{v}_r$. Obviously, if a particle with diameter of d_p is deposited on the surface of the cylinder, the radius is

$$r=a+(d_p/2)=\frac{d_f+d_p}{2} \tag{3.86}$$

Actually, the particle motion is mainly controlled by the radial velocity v_r. From equation (3.79), at $\theta=0$, v_r goes up to maximum, while at $\theta=\pi/2$, and $v_r=0$. Then, the behavior of the particle diffusion is that the deposition mass of the particulate on the cylinder is getting less and less as θ increasing. As a conservative estimation, we just consider the diffusion effect of the windward surface of the cylinder in the range of $0\leqslant\theta\leqslant\pi/2$, while neglect the deposition of particles on the leeside surface of the cylinder in the range of $\pi/2<\theta\leqslant\pi$. The average radial velocity v_r at $r=(d_f+d_p)/2$ is given as

$$\begin{aligned}\bar{v}_r&=\frac{2}{\pi}\int_0^{\pi/2}v_r\mathrm{d}\theta=\frac{2}{\pi}\int_0^{\pi/2}\frac{v_0}{2\mathrm{La}}\left[2\ln\left(\frac{d_f+d_p}{d_f}\right)-1+\left(\frac{d_f}{d_f+d_p}\right)^2\right]\cos\theta\mathrm{d}\theta\\&=\frac{v_0}{\pi\mathrm{La}}\left[2\ln(1+G)-1+\left(\frac{1}{1+G}\right)^2\right]\end{aligned} \tag{3.87}$$

Where G——the interception parameter, $G=d_p/d_f$.

For the problem of particulate diffusion, the particle size is less than 1μm. Then $G=d_p/d_f$ is very small. Therefore, we can expand the part of natural logarithm into series. Equation (3.87) can be simplified as

$$\bar{v}_r\approx\frac{2v_0G}{\pi\mathrm{La}} \tag{3.88}$$

Substituting equation (3.88) into equation (3.85), we have

$$c=c_0\left\{1-\exp\left[-\frac{2\mathrm{Pe}}{\pi d_f\mathrm{La}}(r-a)\right]\right\} \tag{3.89}$$

Where Pe——Peclet number, which is a measure of the relative magnitude of the diffusion motion of the particles and of the convective motion of the air past the cylinder, given by

$$\mathrm{Pe}=\frac{v_0 d_p}{D} \tag{3.90}$$

The mass of the particulate pollutant deposited on the cylinder per square meter and per time can be found by Fick's first law, written as

$$F=-D\left(\frac{\partial c}{\partial r}\right)_{r=\frac{d_f+d_p}{2}} \tag{3.91}$$

The concentrationc in equation (3.91) is given by equation (3.89).

The concentration distribution solution of equation (3.89) is only somewhat crude approximation. When the average radial velocity is used and the particles deposited on the leeside surface of the cylinder is neglected, it will be analogous to the approximation used in the diffusion of a gas flow around a sphere.

3.4.2 Diffusion of a Gas Flow Around a Sphere

The particulate diffusion of a gas flow around a sphere can also be described by Fig. 3.13. Similarly, a viscous flow with low velocity is considered. In the viscous flow condition, the flow function of a gas flow around a sphere is

$$\psi = \frac{1}{2} v_0 \left[1 - \frac{3}{2}\left(\frac{a}{r}\right) + \frac{1}{2}\left(\frac{a}{r}\right)^3 \right] r^2 \sin^2\theta \tag{3.92}$$

The radial and tangential velocity components are

$$v_r = v_0 \left(1 - \frac{3}{2}\frac{a}{r} + \frac{1}{2}\frac{a^3}{r^3} \right) \cos\theta \tag{3.93}$$

$$v_\theta = -v_0 \left(1 - \frac{3}{4}\frac{a}{r} + \frac{1}{4}\frac{a^3}{r^3} \right) \sin\theta \tag{3.94}$$

Here, only the diffusionin normal (radial) direction of the surface of the sphere is discussed. Then, the diffusion equation in polar coordinates is simplified as

$$D\left(\frac{\partial^2 c}{\partial r^2} + \frac{2}{r}\frac{\partial c}{\partial r} \right) + \frac{\partial (v_r c)}{\partial r} = 0 \tag{3.95}$$

Because of $\frac{\partial^2 c}{\partial r^2} >> \frac{2}{r}\frac{\partial c}{\partial r}$, when the idea of the average radial velocity is also used, equation (3.95) becomes equation (3.83). The boundary condition of equation (3.95) is the same as equation (3.84). Then, the particulate diffusion of a gas flow around a sphere can also be expressed by equation (3.85). However, the average radial velocity around the sphere at $r = (d_s + d_p)/2$ is

$$\begin{aligned} \bar{v}_r &= \frac{2}{\pi}\int_0^{\pi/2} v_r \mathrm{d}\theta = \frac{2}{\pi}\int_0^{\pi/2} v_0 \left[1 - \frac{3}{2}\frac{d_s}{d_s + d_p} + \frac{1}{2}\frac{d_s^3}{(d_s + d_p)^3} \right] \cos\theta \mathrm{d}\theta \\ &= \frac{v_0}{\pi}\left[1 - \frac{3}{2(1+G)} + \frac{1}{2}\left(\frac{1}{1+G}\right)^3 \right] \end{aligned} \tag{3.96}$$

Where G ——interception parameter, $G = d_p/d_c$;

d_s ——sphere diameter in m, $d_s = 2a$.

The mass of the particulate pollutant deposited on the sphere per square meter and per time is calculated by Fick' s first law which is written as

$$F = -D\left(\frac{\partial c}{\partial r}\right)_{r = \frac{d_s + d_p}{2}} \tag{3.97}$$

The concentration c in equation (3.97) is given by equation (3.85).

In this chapter, the diffusions of the particulate pollutant on the axisymmetric objects under the condition of the still gas, the steady state flow, and the steady flow around a body have been discussed. The simplification of diffusion model is a idealized approximation processing method.

In the non-axisymmetric cases, the simplification is difficult. Furthermore, the assumption of the absorption wall is used. In practice, there is almost no perfect absorption wall. And also, the effect of absorption materials will be weaken with time increasing. Therefore, the deviations of the theoretical results derived from the assumption of the absorption wall are present. However, it makes sense that the calculation formulae had provided the trend of the concentration variation. Then, the concentration distribution can be used in practice if a certain modification of a theoretical formula is made from experiment[4].

Exercises

3.1 The particles have a density of $2\times10^3\,kg/m^3$. Plot the curves of the terminal settling velocity and diffusion coefficient of the particle size between the ranges from 0.01μm to 10μm at standard conditions.

3.2 There are two walls without absorption effect. The distance between these two walls is b. The pollutant mass of M is emitted from the surfaces of these two walls at $t=0$, as shown in Fig. 3.14. Establish the concentration distribution between these two wall at any time t, and plot the curves of the distribution schematically (Only the first superposition is required).

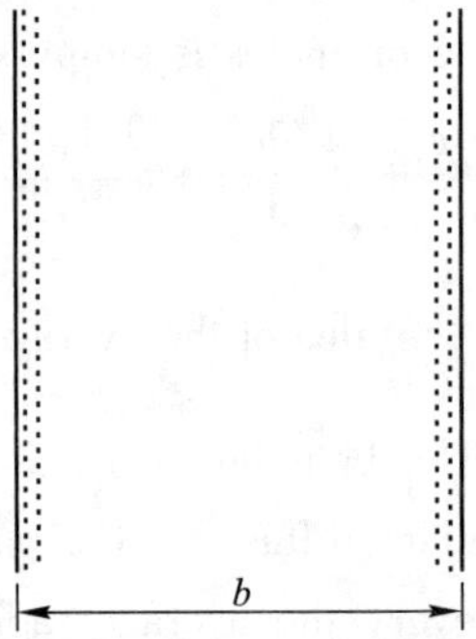

Fig. 3.14 Pollutant emitted from two infinite wall surfaces

3.3 The initial pollutant concentration between two infinite absorption wall is c_0, The distance between two walls is a, Determine the mass of the pollutant which have deposited on per square meter of the absorption wall after time t.

3.4 There are two point pollutant sources with emission strength of M located at the axial line of a cylindrical duct in radius of R, as shown in Fig. 3.15. The distance of these two point sources is x_0, and the duct wall is a reflection wall. Establish the concentration distribution after the second point source ($x>x_0$).

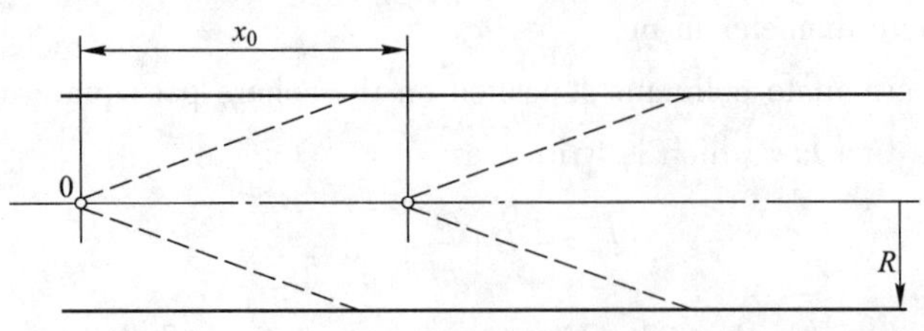

Fig. 3.15 Diffusion of two point source at the axial line of a cylindrical duct

3.5 A cylindrical tube of 0.1m in radius and 0.2m in length is placed in a still aerosol. The particle number concentration of the aerosol is $10^7/m^3$. The inside and outside walls of the tube are absorption walls. Calculate the number of particles deposited on the tube walls after 3s.

3.6 A cylindrical fiber of a in radius and 1.0 m in length is placed vertically in the gas stream with velocity of v_0 as shown in Fig. 3.13. The particle concentration in gas is c_0. Predict the mass of particles collected by this fiber in one second according to the diffusion theory of a gas flow around a cylinder.

References

[1] Friedlander S K. *Smoke, Dust and Haze-Fundamentals of Aerosol Behavior* [M]. Wiley, New York, 1977.

[2] Williams M M R, Loyalka S K. *Aerosol Science Theory and Practice* [M]. Pergamon Press, New York, 1991.

[3] Murray R S. *Advanced Mathematics for Engineers and Scientists* [M]. New York, McGraw-Hill Book Company, 1971.

[4] Seshadri V, Singh S, Kaushal D R, et al. A model for the prediction of concentration and particle size distribution for the flow of multisized particulate suspensions through closed ducts and open channels [J]. Particulate Science and Technology, 2006, 24 (2): 239-258.

4 Coagulation of Aerosol Particulate

The term coagulation is used to describe the process of adhesion or fusion of two particles which takes place when they touch. Such a collision takes place because of the relative velocity between the particles. This relative velocity can arise from a variety of physical causes. It is very useful to discuss the particle coagulation mechanism to separate the fine or super fine particles indirectly. Airborne particulate can collide as a result of Brownian motion, hydrodynamic or electrical forces. If the coagulation is caused by Brownian motion, it is sometime called thermal coagulation[1]. If the coagulation is caused by external force, it is called dynamic coagulation[2]. The dynamic coagulation includes gravitational coagulation, turbulent coagulation, acoustic coagulation, and electrical or electrostatic coagulation. In airborne particulate separation technology, the effect of gravitational coagulation is low. Only when other coagulation method is used in the pollutant gas flow, does hydrodynamic coagulation play a auxiliary role. In the acoustic coagulation, the coagulation rate is quite high. It was once to be thought a hopeful effective method to separate the fine particles. Some practices had proved that the energy consumption was very high. So that we will just discuss Brownian and electrostatic coagulation in this text book.

It is clear that particle coagulation leads to a decrease in the number of particles and an increase in their size (diameter). Therefore, the coagulation rate and the particle size distribution are the key problems which are needed to be solved.

4.1 Brownian Coagulation

4.1.1 Brownian Coagulation of Monodisperse Particles

Here we will look at Brownian coagulation resulting from the relative movement of the particles. A simple case will be examined for spherical and monodisperse particles. It is assumed that particles coagulate after each collision and that the initial change in particle size is small. Furthermore, focus on a single particle, called central particle (which does not undergo Brownian motion), and consider the diffusive flux of the other particles to its surface.

The coagulation rate of monodisperse particulate was expressed by Smoluchowski[3], given by

$$\frac{\mathrm{d}N}{\mathrm{d}t} = -KN^2 \tag{4.1}$$

Where $N(t)$ ——the particle number concentration at time t in/m^3;

K——the coagulation coefficient in m^3/s.

When the particle diameter is larger than the mean free path of gas, K is given by

$$K = 4\pi d_p D = \frac{2k_B T C_u}{3\mu} \quad d_p > \lambda \tag{4.2}$$

Where D——the diffusion coefficient in m^2/s;

k_B——Boltzmann constant, $k_B = 1.38 \times 10^{-23}$ J/K;

T——the absolute temperature in K;

μ——the kinematic viscosity in Pa · s;

C_c——the Cunningham slip correction factor, given by equation (1.3).

It is more useful to know the number concentration at time t. From equation (4.1), we obtain

$$N(t) = \frac{N_0}{1 + N_0 K t} \tag{4.3}$$

Where N_0——the initial number concentration.

Equation(4.3) is not applicable for concentration variations at distances of approximate one gas molecular mean free path from the particle' s surface. Therefore, its region of validity is restricted to particles larger than 0.1μm. The importance of no continuum effects increases as the diameter decreases, and they become significant for diameters smaller than 0.4μm. A correction for the coagulation coefficient was proposed by Fuchs, written as

$$K_B = \beta K \tag{4.4}$$

Then, in equation (4.3), K must be substituted by K_B, written as

$$N(t) = \frac{N_0}{1 + N_0 K_B t} \tag{4.5}$$

At standard conditions (temperature of 298K and pressure of 100kPa), numerical values for the functions β, K and K_B are given in Table 4.1.

Table 4.1 Coagulation coefficients at standard conditions

diameter d_p/μm	Correction coefficient β	$K/m^3 \cdot s^{-1}$	$K_B/m^3 \cdot s^{-1}$
0.004	0.037	168×10^{-16}	6.2×10^{-16}
0.01	0.14	68×10^{-16}	9.5×10^{-16}
0.04	0.58	19×10^{-16}	10.7×10^{-16}
0.1	0.82	8.7×10^{-16}	7.2×10^{-16}
0.4	0.95	4.2×10^{-16}	4.0×10^{-16}
1	0.97	3.4×10^{-16}	3.4×10^{-16}
4	0.99	3.1×10^{-16}	3.1×10^{-16}
10	0.99	3.0×10^{-16}	3.0×10^{-16}

It is noticed that from Table 4.1, for monodisperse particulate, the coagulation coefficient K_B is less dependent particle size to a certain extent. In steady state, the coagulation coefficient is approximately taken as constant. The coagulation rate is only proportional to the square of the particle number concentration at time t. When the number concentration is high, the process of the coagulation is fast. While the number concentration is low, the process of the coagulation is slow.

Then, how about the particle size changing with the number decrease? In an enclosed system without loss, the mass of the particulate keeps constant. If a half of particle number is decreased, the mass of every particle is doubled. For the spherical particle, the particle size is proportional to the third root of the particle volume, and also proportional to the third root of the number concentration, written as

$$d(t) = d_0\left[\frac{N_0}{N(t)}\right]^{\frac{1}{3}} \tag{4.6}$$

Where d_0——the initial particle size at $t=0$.

Therefore, if the number concentration is reduced to 1/8 of initial concentration, the particle size just become two times of the initial size. Combining equations (4.5) and (4.6), we know the particle size variation at any time due to the coagulation, given by

$$d(t) = d_0(N_0 K_B t)^{\frac{1}{3}} \tag{4.7}$$

Equation (4.7) is suitable for water droplets and almost suitable for compact solid particles. The times of a half of reduction of number concentration and doubled increment of particle size are given in Table 4.2. We can see clearly that Brownian coagulation rate is very slow.

Table 4.2 Initial concentration and time of coagulation①

Initial concentration N_0/m^{-3}	Times of a half of reduction of number concentration /s	Times of doubled particle size /s
10^{18}	0.002	0.014
10^{16}	0.2	1.4
10^{14}	20	140
10^{12}	2 000 (33min)	14 000 (4h)
10^{10}	200 000 (55h)	1 400 000 (16d)

① Coagulation coefficient of monodisperse particulate is $K=5\times10^{-16}m^3/s$.

4.1.2 Brownian Coagulation of Polydisperse Particles

In general, the primary particles that compose an aggregate are polydisperse. The problem becomes complex. However, for implicity, and in the absence of appropriate theoretical models, we will analyse only aggregates composed of identical spherical monomers.

The coagulation of polydisperse particulate is different from that of monodisperse particulate because the coagulation rate is decided by the diffusion from a particle to others. When a fine particle with high diffusion coefficient diffuses to a large particle, the coagulation rate is enhanced. For example, if the size difference between two particles is 10 times, coagulation rate is accelerated to 3 times, and if the size difference between two particles is 100 times, coagulation rate is accelerated to over 25 times.

Suppose the number concentration of a polydisperse particulate is n, a particle with radius r_1

collides a central particle with radius r_2. Then the reduced numbers of all particles with radius r_1 are equal to the collision times between these two size particles, given by

$$\frac{dn}{dt} = -4\pi r D_1 r_1 \left(\frac{n^2}{2}\right) \tag{4.8}$$

Where D_1——the diffusion coefficient of the particle with radius r_1 in m^2/s.

To avert twice calculation of the coagulation in equation (4.8), n^2 is divided by 2.

Because the diffusion is mutual. In equation (4.8), D_1 and r_1 will be substituted by $D_1 + D_2$ and $r_1 + r_2$ respectively. Then

$$\frac{dn}{dt} = -2\pi(r_1 + r_2)(D_1 + D_2)n^2 \tag{4.9}$$

From equation (3.2), diffusion coefficient is rewritten as

$$D = k_B T C_c / 6\pi\mu r \tag{4.10}$$

Substituting equation (4.10) into equation (4.9), a more general Brownian coagulation equation is developed by

$$\frac{dn}{dt} = -\frac{K}{2}n^2 \tag{4.11}$$

Equation (4.11) is analogous to equation(4.2), but

$$K = \frac{2k_B T}{3\mu}\left[\frac{C_c(r_1)}{r_1} + \frac{C_c(r_2)}{r_2}\right](r_1 + r_2) \tag{4.12}$$

The solution of equation (4.11) is

$$n = n_0 \Big/ \left(1 + \frac{1}{2}Kn_0 t\right) \tag{4.13}$$

Where n_0——the initial number concentration at $t=0$ in $/m^3$.

Equation (4.11) or equation(4.13) is typical Brownian coagulation or thermal coagulation formula because it is the fundamentals of other coagulation mechanisms. No matter what kind of coagulation, the coagulation equation is the same exception of the coagulation coefficient. The coagulation coefficient is the characterization of the coagulation rate. Therefore, one of the most important research works in the coagulation of airborne particulate is to find the coagulation coefficient K.

The coagulation coefficient K calculated by diffusion theory and air kinetic theory is given in Table 4.3. A distinguishing feature has been found that the coagulation rate of the same size particles is much slower than that of the different size particles. That is, the larger the geometric standard deviation is, the more favorable the coagulation.

From Table 4.3, it can be seen that, for polydisperse particulate, the range of coagulation coefficient value is wide. It is not convenient to analyze quantitatively. Lee and Chen had proposed an empirical calculation formula of average coagulation coefficient $\overline{K}$ for the log-normal distribution of particulate[4]. The number median diameter (NMD) and the geometric standard deviation σ_g

Table 4.3 Coagulation coefficient K (Cunningham slip correction factor C_c is considered)

$r_1/\mu m$	$K/m^3 \cdot s^{-1}$			
0.001	87.8×10^{-16}	—	—	—
0.01	180.2×10^{-16}	21.0×10^{-16}	—	—
0.1	8845×10^{-16}	168.5×10^{-16}	11.10×10^{-16}	—
1.0	178100×10^{-16}	2032×10^{-16}	35.95×10^{-16}	6.44×10^{-16}
$r_2/\mu m$	0.001	0.01	0.1	1.0

are included in this formula

$$\overline{K} = \frac{2k_B T}{3\eta}\left\{1 + \exp(\ln^2\sigma_g) + \left(\frac{2.49\lambda}{\text{NMD}}\right)\left[\exp(0.5\ln^2\sigma_g) + \exp(2.5\ln^2\sigma_g)\right]\right\} \quad (4.14)$$

Only NMD is changing appropriately, can K be substituted by $\overline{K}$ in equation (4.13) to predict the concentration during time t. The NMD can be calculated by equation (4.14) during time t. $\overline{K}$ is thought to be constant during this time. In this calculation, the geometric standard deviation σ_g is assumed constant either. If the particle size is changing greatly, the calculation should be separated several steps. The value of $\overline{K}$ is different in each step.

Example 4.1 (1) At standard condition, the initial concentration of ferric oxide particles in gas is $10^{13}/m^3$. The monodisperse particulate with particle size 0.2μm is assumed. Calculate the number concentration and the coagulated particle size after two min.

(2) At standard condition, the geometric standard deviation σ_g of the polydisperse particulate is constant. NMD is 0.2μm, and σ_g is 2. Calculate the number concentration and the coagulated particle number mediam diameter (NMD) after 2min.

Solution

(1) Using equation (4.3), it is writen as

$$N(t) = \frac{N_0}{1 + N_0 K t}$$

where, $N_0 = 10^{13}/m^3$, and

$$K = \frac{4k_B T C_u}{3\mu} = 5.6 \times 10^{-16}\ m^3/s$$

At $t = 120s$, then

$$N(t) = 5.98 \times 10^{12}\ /m^3$$

According to equation (4.6), the particle size after 2min is

$$d(t) = d_0\left(\frac{N_0}{N(t)}\right)^{\frac{1}{3}} = 0.24\ \mu m$$

(2) Use equation (4.14) to calculate the average coaqulation coefficient as

$$\bar{K} = \frac{2kT}{3\mu}\left\{1 + \exp(\ln^2\sigma_g) + \frac{2.49\lambda}{\text{NMD}}[\exp(0.5\ln^2\sigma_g) + \exp(2.5\ln^2\sigma_g)]\right\}$$

$$= 1.04 \times 10^{-15}\ \text{m}^3/\text{s}$$

Substitute the value of $\bar{K}$ to K in equation (4.13), the number concentration is

$$N(t) = \frac{N_0}{1 + N_0\bar{K}t} = 4.45 \times 10^{12}\ /\text{m}^3$$

According to equation (4.6), the coagulated particle number median diameter (NMD) after two min can be calculated as

$$d(t) = \text{NMD} = d_0\left(\frac{N_0}{N(t)}\right)^{\frac{1}{3}} = 0.26\ \mu\text{m}$$

4.2 Electrical Coagulation

Electrostatic coagulation is one of the most possibly coagulation methods of separating the fine particles from the pollutant gas since thecoagulation coefficient is much higher than Brownian coagulation. Electrostatic coagulation includes Coulomb's coagulation and bipolar charged particulate coagulation in alternating electric field.

4.2.1 Coulomb's Coagulation

It is clear that charge will influence the way in which particles interact with each other. For example, unipolar charging (like charges) will surely lead to a reduced interaction and hence less coagulation. On the other hand, bipolar charging (unlike charges) will lead to enhance coagulation[5]. Here, only the coagulation of bipolar charged particles in gas is considered.

As we have known that the coagulation rate depends on the coagulation coefficient. Even in the case of bipolar charged particles, actually, no matter it is symmetrical or not, the attractive and repel forces are present at same time. We just know that in symmetrical bipolar system, the attractive force dominates the coagulation processes. Then the repel force effect is negligible. Coulomb's coagulation coefficient in bipolar charged particulate system was given by Williams and Loyalka[6] as

$$K_c = \frac{z(e^z + 1)}{2(e^z - 1)}K \tag{4.15}$$

where,

$$z = \frac{q_1 q_2}{2\pi k_B T\varepsilon_0(x_1 + x_2)} \tag{4.16}$$

Where x_1, x_2——the diameters of two particles;

q_1, q_2——the charges on each particle;

ε_0——the vacuum permittivity, $\varepsilon_0 = 8.85 \times 10^{-12}$ C/(V · m).

If the particles are symmetrically bipolar charged, equation (4.15) may be more reasonable. If not, it is somewhat over estimated.

Example 4.2 A bipolar monodisperse aerosol particulate is charged symmetrically in the electric field strength of $E = 4\text{kV/cm}$. The relative permittivity of the particle is $\varepsilon = 6$. Calculate the Coulomb's coagulation coefficient of saturated charged particle with diameter of 1μm.

Solution

A particle can be charged in electric field by field charging and diffusion charging. If the particle size is larger than 1μm, the diffusion charging is negligible. If the particle diameter is smaller than 1μm, both field charging and diffusion charging are need to be considered. A combined formula of field charging and diffusion charging was given by Cochet (shown in Chapter 7) as

$$q = \pi\varepsilon_0 E d_p^2\left[\left(\frac{\varepsilon - 1}{\varepsilon + 2}\right)\left(\frac{2}{1 + 2\lambda/d_p}\right) + (1 + 2\lambda/d_p)^2\right] = 1.245 \times 10^{-17}\text{C}$$

From equation (4.16), it is written as

$$z = \frac{q_1 q_2}{2\pi k_B T\varepsilon_0(x_1 + x_2)} = 3.45 \times 10^2$$

Using equation (4.2), Brownian coagulation coefficient of monodisperse particles is

$$K = 4\pi d_p D = \frac{2k_B T C_u}{3\mu} = 1.5 \times 10^{-16}\ \text{m}^3/\text{s}$$

Then, according to equation (4.15), Coulomb's coagulation coefficient is

$$K_c = \frac{z(e^z + 1)}{2(e^z - 1)}K \approx 2.6 \times 10^{-14}\ \text{m}^3/\text{s}$$

For 1μm particle, the range of Coulomb's coagulation coefficient is from $10^{-13} \sim 10^{-14}\text{m}^3/\text{s}$ in this example. The result is in agreement with Gutsch and Loffler[7]. The Coulomb's coagulation coefficient is about 10^2 times greater than Brownian coagulation coefficient[8].

Some approaches of submicron particle collection had been made on Coulomb's coagulation of bipolar charged particulate[9~13]. A basic principle of collection fine particles by Coulomb's coagulation is illustrated in Fig. 4.1.

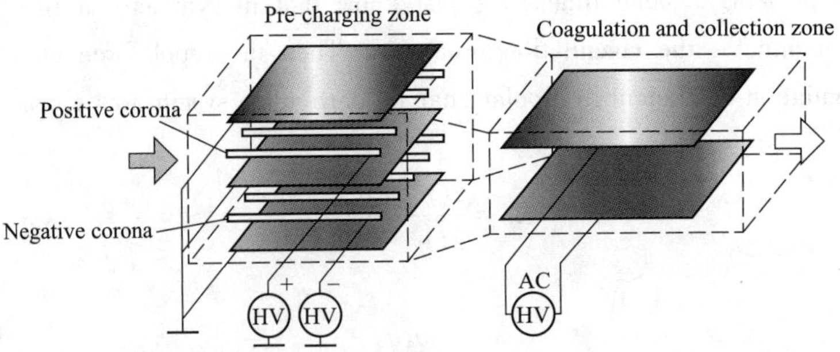

Fig. 4.1 Bipolar charged particulate collection by Coulomb's coagulation

Particles are bipolar charged by the positive and negative corona discharging in pre-charging zone. Then, the bipolar charged particles go into the coagulation zone. The bipolar changed particles attracts each other due to Coulomb's force to form the larger particles. At last, the coagulated particles are collected in the electric field of the collection zone.

Even though Coulomb's coagulation coefficient is much greater than Brownian coagulation coefficient, the coagulation rate of Coulomb's coagulation is not high enough for engineering application[14]. Thus, if we want to accelerate the coagulation rate, it is necessary to introduce the external force field into the electrical coagulation.

4.2.2 Electrostatic Coagulation in an Alternating Electric Field

The coagulation rate of bipolar charged particles in an alternating electric field had been thought to be higher than that of other coagulation methods[15]. A typical apparatus of fine particle collection by electrostatic coagulation in an alternating electric field is illustrated schematically in Fig. 4.2. There are also three zones. First, the bipolar particles are formed in the bipolar charging zone. Then, these bipolar particles are coagulated in coagulation zone with an alternating electric field. Finally, the coagulated particles are collected by the collection zone which is an electrostatic precipitator.

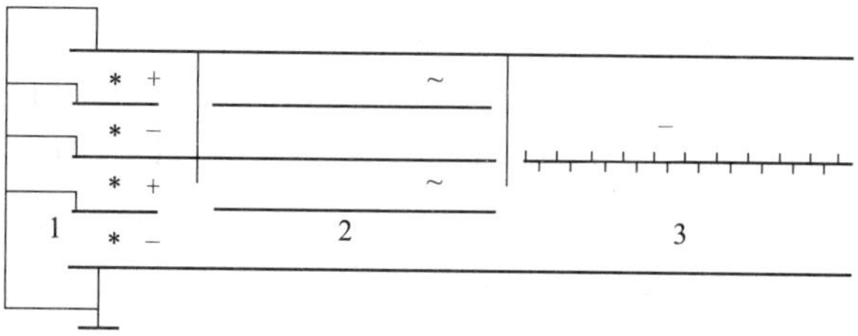

Fig. 4.2 Bipolar charged particulate collection by electrostatic coagulation in an alternating electric field
1—Pre-charging zone; 2—Coagulation zone with an alternating electric field; 3—Collection zone

It is clear that under the alternating electric force, the bipolar charged particles are vibrating in the alternating electric field. As a result, the possibility of the collision of the bipolar charged particles is increased. According to the coagulation coefficient formula developed by Kildes et. al.[16], for 1μm particle, at alternating electric field strength $E = 4\text{kV/cm}$, the predicted value of the coagulation coefficient of bipolar charged particles is $10^{-12}\text{m}^3/\text{s}$. While Coulomb's coagulation coefficient is in the range from $10^{-13} \sim 10^{-14}\text{m}^3/\text{s}$.

In an external force field, the coagulation of the collision of two particles depends on the relative velocities of these two particles, i. e., the difference of the velocity. Suppose two particles with diameters of x_1 and x_2, and with velocities of v_1 and v_2. The coagulation coefficient is

$$K = \frac{\pi}{4}(x_1 + x_2)^2 |\vec{v}_1 - \vec{v}_2| \tag{4.17}$$

In the alternating electric field, the vibrating velocity of a charged particle is determined by equation (2.44). Suppose the charge on a particle with diameter of x_1 is positive and another particle with diameter of x_2 is negative, the vibrating velocity of the positive and negative charged particles x_1 and x_2 are

$$\vec{v}_1 = \frac{\tau_1 q_1 \vec{E}_0}{m_1}\sin\psi t, \quad \vec{v}_2 = -\frac{\tau_2 q_2 \vec{E}_0}{m_2}\sin\psi t \tag{4.18}$$

Because of the particle velocity being the function of time, the coagulation coefficient of charged particles in alternating electric field K_ψ is also the function of time. The question become more complex. As an approximate prediction, it is noticed that the vibrating period $2\pi/\psi$ is short, the method of finding acoustic coagulation coefficient proposed by Williams and Loyalka[6] is used. That is, the average coagulation coefficient value in a half period is taken. In this way, the coagulation coefficient of charged particles in alternating electric field K_ψ can be described by a integral between 0 and π/ψ as

$$K_\psi = \frac{\pi}{4}(x_1 + x_2)^2 \frac{\psi}{\pi}\int_0^{\pi/\psi} |v_1 - v_2| dt \tag{4.19}$$

Substituting equation (4.18) into equation (4.19), we obtain

$$K_\psi = \frac{(x_1 + x_2)^2 E_0}{6\pi\mu}\left(\frac{q_1}{x_1} + \frac{q_2}{x_2}\right) \tag{4.20}$$

When the inertial effect is neglected, the coagulation coefficient of charged particles in alternating electric field is not related with the radian frequency ψ. However, because a particle always possesses inertia, the radian frequency ψ, of course, will affect the coagulation behavior of the changed particles. If the radian frequency is too high, the amplitude of vibration is getting very short. And then, the coagulation coefficient becomes smaller. The experimental results of Watanabe indicated that the coagulation coefficient can get to the maximum when the frequency of alternating electric field is about 10Hz or less[17] when the bipolar charged particles are coagulated in the alternating electric field.

4.3 Particle Distribution in Coagulation Process

The airborne particulate is polydisperse. After the particles are coagulated, the particle size is still polydisperse. If we discuss the particle size distribution in the process of the coagulation of different size particles at the very start, we have to solve the dynamic equation numerically which is a troublesome work and less applicable in practice.

According to the coagulation equation (4.11), as soon as the coagulation coefficient is found, the particle number concentration reduction with time can be calculated. However, this concentration just indicates the information of number of the particles at any time. The main purpose of discussing the coagulation is to know the particle size distribution after aperiod of time t. Therefore, in this section, an approximate theoretical particle size distribution function of the coagulated particles will be described following.

4.3.1 Assumption of Self-preserving

It is of considerable practical interest to know whether the size distribution of the initial aerosol particles affects the size distribution of the coagulated particles after some time. In order to determine the size distribution of the coagulated particles, the assumption of self-preserving was proposed by Friederlander[18].

The size distribution of the particulate coagulation can be solved using a similarity transformation. Such solutions represent asymptotic forms approached after long time and are independent of the initial size distribution. In carrying out the similarity transformation, it is assumed that the fraction of the particle volume normalized by the average particle volume as

$$\frac{f_n(v, t)\mathrm{d}v}{N(t)} = \psi\left(\frac{v}{\bar{v}}\right)\mathrm{d}\left(\frac{v}{\bar{v}}\right) \tag{4.21}$$

Where $f_n(v, t)$——the function of the volume;

$\bar{v}$——the average particle volume at time t, $\bar{v} = V(t)/N(t)$.

Rearranging the equation (4.21), the dimensional size distribution function can be expressed as

$$n(v, t) = \frac{N(t)^2}{V(t)}\psi(\eta) \tag{4.22}$$

where,

$$\eta = v/\bar{v} = vN(t)/V(t)$$

Note that

$$N(t) = \int_0^\infty n(v, t)\mathrm{d}v \tag{4.23a}$$

$$V(t) = \int_0^\infty vn(v, t)\mathrm{d}v \tag{4.23b}$$

Using the number frequency of particle diameter function (distribution function) $f_n(x, t)$ to express, the similarity transformation becomes

$$f_n(x, t) = \frac{N^{4/3}}{V^{1/3}}\psi(\eta_x) \tag{4.24}$$

Where x——the particle diameter at time t.

Here, for discussion convenience, we use x instead of d_p to indicate particle diameter, $\eta_x = x(N/V)^{1/3}$.

The size distribution is thus represented in terms of N and V and a dimensionless function $\psi(\eta_x)$. The shapes of the size distribution are assumed to be similar at different times, multiplied by the time-dependent factor N^2/V. These solutions are called self-preserving.

N and V are the ordinary functions of time t. In the simple case, the case of pure coagulation with on sources and sinks, the V is constant. But due to the coagulation, the number of particles N is reducing. If we knew the initial particle number, N can be found at any time t according the coagulation equation [such as Brownian coagulation equation (4.13)].

To determine $\psi(\eta)$, the relation (4.22) for $n(v, t)$ is substituted into general dynamic equation, an ordinary integro-differential equation for $\psi(\eta)$ is obtained. The general dynamic equation is given by

$$\frac{\partial n(v, t)}{\partial t} = \frac{1}{2}\int_0^v K(u, v-u)n(v-u, t)n(u, t)\mathrm{d}u - n(v, t)\int_0^\infty K(u, v)n(u, t)\mathrm{d}u \tag{4.25}$$

For particle diameter $x > 1\mu\mathrm{m}$ (continuum regime), the Brownian coagulation coefficient is given by

$$K(u,\ v)=\frac{2k_{\mathrm{B}}\mathrm{T}}{3\mu}\left[2+\left(\frac{u}{v}\right)^{1/3}+\left(\frac{v}{u}\right)^{1/3}\right] \tag{4.26}$$

For particle diameter $x \leqslant 1\mu\mathrm{m}$ (transition regime), the Brownian coagulation coefficient is given by

$$K(u,\ v)=\left(\frac{8\pi k_{\mathrm{B}}T}{\rho}\right)^{1/2}\left(\frac{3}{4\pi}\right)^{2/3}(u^{1/3}+v^{1/3})^2\left(\frac{1}{u}+\frac{1}{v}\right)^{1/2} \tag{4.27}$$

The function $\psi(\eta_x)$ has been solved numerically from equation (4.25) by Friedlander and Wang[19].

4.3.2 Particle Size Distribution Simplification in Coagulation Process

It can be seen that, from above discussion, based on theassumption to determine the particle size distribution of the coagulated particles is very complicated. However, the idea of that the shapes of the size distribution are similar at different times is very useful in the assumption of self-preserving.

Theory and practice of aerosol science and technology have been shown that the size distributions of almost all of the particulates produced by human or nature follow the log-normal distribution. It is deduced that the population of the coagulated particles ought to follow the log-normal distribution either. In the process of the particle coagulation, the particle size is getting larger, and the number is getting less. But in order to ensure that the shapes of the size distribution are similar at different times, the geometric standard deviation of the coagulated particles should be kept no change. Vemury et. al[20] had proved experimentally that whether the particles are charged or not, the variation of geometric standard deviation is very little.

The initial particle number concentration has been known already as

$$f_n(x,\ 0)=\frac{1}{x\sqrt{2\pi}\ln\sigma}\exp\left[-\left(\frac{\ln x-\ln a_0}{\sqrt{2}\ln\sigma}\right)^2\right] \tag{4.28}$$

Where a_0——initial number median diameter of the particles in m;

σ——geometric standard deviation of the particles in m;

x——particle diameter in m.

Therefore, the particle number distribution can be written directly after a period of coagulation time t as

$$f_n(x,\ t)=\frac{1}{x\sqrt{2\pi}\ln\sigma}\exp\left[-\left(\frac{\ln x-\ln a(t)}{\sqrt{2}\ln\sigma}\right)^2\right] \tag{4.29}$$

Where $a(t)$ ——the number median diameter of the coagulated particles in m.

Equation (4.29) is the function of time because the number median diameter of the coagulated particles a is the function of time. The next work is to determine the number median diameter of the coagulated particles at time t.

The mass concentration of the initial particles in gas is

$$C = V_0 \rho_x \tag{4.30}$$

Where V_0——total volume of particles per cubic meter of gas in m^3/m^3;

ρ_x——particle density in kg/m^3.

For the initial number concentration n_0 in gas, it is easy to prove that (Shown in Example 1.3)

$$V_0 = \frac{\pi}{6} n_0 \int_0^{\infty} f(x) x^3 dx = \frac{\pi}{6} n_0 \exp(3\ln a_0 + 4.5\ln^2\sigma) \tag{4.31}$$

Substituting equation (4.31) into equation (4.30), yields

$$C = \frac{\pi}{6} n_0 \rho_x \exp(3\ln a_0 + 4.5\ln^2\sigma) \tag{4.32}$$

After time t, the number concentration n_0 has been reduced to $n(t)$, while the particle size has been growing up from a_0 to $a(t)$. Similarly, the mass concentration is given by

$$C = \frac{\pi}{6} n(t) \rho_x \exp(3\ln a(t) + 4.5\ln^2\sigma) \tag{4.33}$$

Note the mass concentration is constant during coagulation. Therefore, equation (4.32) must be equal to equation (4.33). Then, we obtain

$$n(t)/n_0 = [a_0/a(t)]^3 \tag{4.34}$$

Substituting equation (4.34) into equation (4.13), the number median diameter of the coagulated particles at time t is written as

$$a(t) = (1 + Kn_0 t/2)^{1/3} a_0 \tag{4.35}$$

Substituting equation (4.35) into equation (4.29), the number particle size distribution after a period of coagulation time is

$$f_n(x, t) = \frac{1}{x\sqrt{2\pi}\ln\sigma} \exp\left[-\left(\frac{\ln x - \ln(1 + Kn_0 t/2)^{1/3} a_0}{\sqrt{2}\ln\sigma}\right)^2\right] \tag{4.36}$$

In coagulation research, the attention has been paid to the variation of the particle number concentration. However, in the industrial application, the mass concentration is commonly used. Usually, the initial particle mass concentration C_0, mass median diameter d_0, and particle density ρ_x are already known. If the coagulated particle mass median diameter $d(t)$ is found, the particle mass particle size distribution can be obtained.

According to the relation between the number and mass median diameter, it is given by

$$d_0 = a_0 \exp(3\ln^2\sigma), \quad d(t) = a(t)\exp(3\ln^2\sigma) \tag{4.37}$$

Substituting equation (4.35) into (4.37), we obtain

$$d(t) = (1 + Kn_0 t/2)^{1/3} a_0 \exp(3\ln^2\sigma) \tag{4.38}$$

Using equation (4.30), we have

$$C_0 = \frac{\pi}{6} n_0 a_0^3 \rho_x \quad \text{or} \quad n_0 = C_0 \Big/ \left(\frac{\pi}{6} a_0^3 \rho_x\right) \tag{4.39}$$

Substituting equation (4.39) into equation (4.38), then the coagulated particle massmedian diameter $d(t)$ is expressed as

$$d(t) = \left[1 + \frac{3KC_0}{\pi a_0^3 \rho_x} t\right]^{1/3} [\exp(3\ln^2\sigma)] a_0 = \left[1 + \frac{3KC_0}{\pi a_0^3 \rho_x} t\right]^{1/3} d_0 \tag{4.40}$$

Now, the mass particle size distribution function can be written directly as

$$f_m(x,\ t) = \frac{1}{x\sqrt{2\pi}\ln\sigma}\exp\left[-\left(\frac{\ln x - \ln d(t)}{\sqrt{2}\ln\sigma}\right)^2\right] \tag{4.41}$$

Thus, if the number median diameter a_0 or the mass median diameter d_0, the geometric standard deviation σ, and the initial particle concentration n_0 are known, the number particle size distribution and the mass particle size distribution function of the coagulated particles at any time can be obtained by equations (4.36) and (4.41).

We can see that the particle size distribution developed by means of the assumption of unchanging geometric standard deviation is very simple.

In the assumption of unchanging geometric standard deviation, the coagulation coefficient is determined by the initial particle number median diameter a_0. In other words, the polydisperse particulate is taken as the monodisperse particulate with particle diameter a_0. It is obviously not precise. In this case, the coagulation coefficient is underestimated.

Example 4.3 The initial particle number concentration of a polydisperse particulate is $n_0 = 1 \times 10^{12}/\text{m}^3$. The number median diameter is $a_0 = 0.1\mu\text{m}$. The geometric standard deviation is $\sigma = 1.5$. Determine the particle size distribution of this particulate undergoing the Brownian coagulation after 3h.

Solution

For the polydisperse particulate Brownian coagulation and the particle diameter of 0.1μm, as shown in Table 4.3, the coagulation coefficient is

$$K = 11.1 \times 10^{-16}\ \text{m}^3/\text{s}$$

3h later, the number median diameter of the coagulated particles is calculated by equation (4.35) as

$$a = (1 + Kn_0t/2)^{1/3}a_0 = 0.19\ \mu\text{m}$$

According to equation (4.36), the number particle size distribution after 3h is

$$f_n(x,\ t) = \frac{1}{x\sqrt{2\pi}\ln 1.5}\exp\left[-\left(\frac{\ln x - \ln 0.19}{\sqrt{2}\ln 1.5}\right)^2\right]$$

Exercises

4.1 The number concentration of the monodisperse cupric oxide (CuO) aerosol at standard conditions is $10^{12}/\text{m}^3$, and the initial cupric oxide particle diameter is 0.1μm. Calculate the time of a half of the concentration reduction.

4.2 Suppose a bipolar aerosol is monodisperse and symmetrical. Calculate the coagulation coefficient of a 5μm particle with the charge of 2×10^{-16} C in this bipolar charged aerosol.

4.3 The initial particle number concentration of a polydisperse particulate is $3\times10^{13}/\text{m}^3$, the number median diameter is $a_0 = 1\mu\text{m}$, the geometric standard deviation is $\sigma = 1.5$, and the coagulation coefficient is $K = 10^{-14}\text{m}^3/\text{s}$. Calculate the particle number concentration of the polydisperse particulate after 1h.

4.4 A bipolar aerosol in the alternating electric field is monodisperse and symmetrical, the charge on a particle

1μm in diameter is 5×10^{-17}C. and the electric field strength follows $E = 4 \times 10^5 \sin(\pi/2)t$. Predict the average coagulation coefficient K_ω.

4.5 It is known that the particle density is 2000kg/m^3, the particle number concentration in the particulate pollutant is 1×10^{13}/m^3, the number median diameter is 0.1μm, the Coulomb's coagulation coefficient is 1×10^{-14} m^3/s, and the geometric standard deviation is 1.5. Predict the number median diameter and mass median diameter of the particulate pollutant after 1h (the self-preserving is assumed in this coagulation process).

References

[1] Otto E, Fissan H. Brownian coagulation of submicron particles [J]. Advanced Powder Technology, 1999, 10 (1): 1-20.

[2] Pratsinis S E, Kim K. Particle coagulation, diffusion and thermophoresis in laminar tube flows [J]. Journal of Aerosol Science, 1989, 20 (1): 101-111.

[3] Yaghouti M R, Rezakhanlou F, Hammond A, et al. Coagulation, diffusion and the continuous Smoluchowski equation [J]. Stochastic Processes and their Applications, 2009, 119 (9): 3042-3080.

[4] Lee K W, Chen H. Coagulation rate of polydisperse particles [J]. Aerosol Science and Technology, 1984, 3 (3): 327-334.

[5] Tan B, Wang L, Wu Z, et al. An approximate expression for the coagulation coefficient of bipolarly charged particles in an alternating electric field [J]. Journal of Aerosol Science, 2008, 39 (9): 793-800.

[6] Williams M M R, Loyalka S K. *Aerosol Science Theory and Practice* [M]. New York, Pergamon Press, 1991.

[7] Gutsch A, Loffer F. Electrically enhanced agglomeration of nanosized aerosol [J]. Journal of Aerosol Science, 1994, 25 (3): 307-308.

[8] Eliasson B, Egli W. Bipolar coagulation—modeling and applications [J]. Journal of Aerosol Science, 1991, 22 (4): 429-440.

[9] Chen D, Wu K, Mi J, et al. Experimental investigation of aerodynamic agglomeration of fine ash particles from a 330 MW PC-fired boiler [J]. Fuel, 2016: 86-93.

[10] Ji J H, Huang J, Bae G, et al. Experimental Study on Electrical Agglomeration of Liquid Particles in an Alternating Electric Field [J]. Transactions of The Korean Society of Mechanical Engineers B, 2001, 25 (3): 442-450.

[11] Chao H, Xiuqin M, Youshan S, et al. Particle agglomeration in bipolar barb agglomerator under AC electric field [J]. Plasma Science & Technology, 2014, 17: 317-320.

[12] Xiang X, Chang Y, Nie Y, et al. Investigation of the performance of bipolar transverse plate ESP in the sintering flue control [J]. Journal of Electrostatics, 2015, 76: 18-23.

[13] Nakajima Y, Sato T. Electrostatic collection of submicron particles with the aid of electrostatic agglomeration promoted by particle vibration [J]. Powder Technology, 2003, 135: 266-284.

[14] Koiznmi Y, et al. Estimation of the agglomeration coefficient of bipolar-charged aerosol particles [J]. Electrostatics, 2000, 48 (2): 9-10.

[15] Lehtinen K E, Jokiniemi J, Kauppinen E I, et al. Kinematic coagulation of charged droplets in an alternating electric field [J]. Aerosol Science and Technology, 1995, 23 (3): 422-430.

[16] Kildeso J, Bhatia V K, Lind L, et al. An experimental investigation for agglomeration of aerosols in alternating electric fields [J]. Aerosol Science and Technology, 1995, 23 (4): 603-610.

[17] Watanabe T, Tochikubo F, Koizurni Y, et al. Submicron particle agglomeration by an electrostatic agglomerator [J]. Journal of Electrostatics, 1995, 34 (4): 367-383.

[18] Friedlander S K. *Smoke, Dust, and Haze: Fundamentals of Aerosol Dynamics*, 2nd [M]. New York: Oxford University Press, 2000.

[19] Friedlander S K, Wang C S. The self-preserving particle size distribution for coagulation by brownian motion [J]. Journal of Colloid and Interface Science, 1966, 22 (2): 126-132.

[20] Vemury S, Janzen C, Pratsinis S E, et al. Coagulation of symmetric and asymmetric bipolar aerosols [J]. Journal of Aerosol Science, 1997, 28 (4): 599-611.

5 Aerodynamic Separation of Particulate

When particulate in a gas stream is acted upon by external forces, the particles acquire a velocity component in a direction different from that of the gas stream. In order to separate the particulate from a gas stream, one must be able to compute the motion of a particle in an external force field. Particulate removal devices rely on one or more of the forces, such as gravitational force, inertial force, electric force, diffusiophoresis force, thermophoresis forces, and magnetic force. The devices, such as settling chamber, centrifugal separator, electrostatic precipitator, filter, and scrubber, are most commonly used for controlling the particulate pollutants[1].

Aerodynamic separation refers to utilize the inertial force to separate the particles from a gas[2]. The mechanism of settling chamber does not belongs to aerodynamic separation. And the collection efficiency is comparatively low. However, the method of the particle separation analysis in settling chamber is very useful for other separation mechanism discussions. Therefore, the settling chamber is described first in this chapter.

5.1 Settling Chamber

Settling chamber can be used to separate out the larger particles in a particulate distribution. The settling chamber is almost the cheapest device to construct, operate, and maintain. The main usefulness of settling chamber, however, lies in serving as a preliminary collection device for a more efficient control device[3].

Settling chamber offers the advantages of (1) low cost, (2) small pressure drop, and (3) collection of particles without need for water. The main disadvantage of settling chamber is the large space that it requires.

5.1.1 Settling Chamber of Laminar Flow

In analyzing the performance of a settling chamber, the key feature is the nature of the gas flow through the device. In the laminar flow settling chamber, we assume that: (1) the particles are introduced uniformly across the entrance to the channel at concentration C_0, (2) the gas velocity u is uniformly distributed in any cross section, and (3) there is no vertical mixing of particles.

Consider the laminar flow settling chamber shown in Fig. 5.1. The particle-laden gas flow in a rectangular chamber with H in height, B in width, and L in length.

According to equation (2.23), the settling velocity of a particle is

$$v_t = \tau g \tag{5.1}$$

If a particle with diameter d_p at height of h of the inlet section is going to the end of the

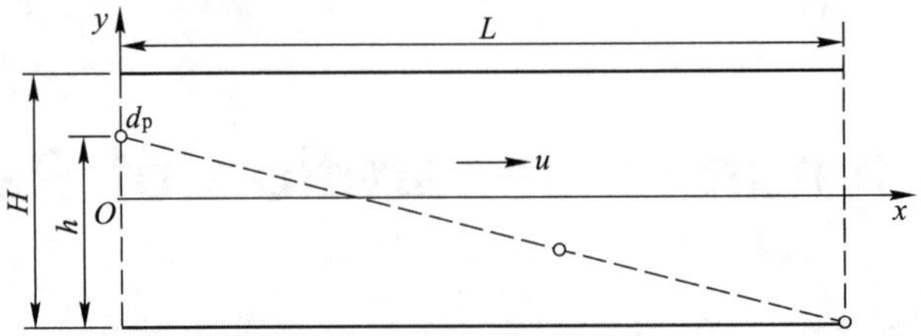

Fig. 5. 1　Gravitational settling of a particle in a horizontal chamber

chamber along the dotted line. This particle just touches the end of the lower plate. We say this particle is captured. It is clear that all the particles with diameter d_p below this dotted line must be captured. Therefore, the fraction of the collected particles, which is called the fractional collection efficiency, is given by

$$\eta_1 = \frac{h}{H} \tag{5.2}$$

Where　h——the particle settling distance, which is given by

$$h = v_t t = v_t \frac{L}{u} \tag{5.3}$$

Substituting equation (5.3) into equation (5.2), the fractional collection efficiency of the laminar flow settling chamber is given by

$$\eta_1 = \frac{v_t L}{uH} \tag{5.4}$$

If the velocity distribution is considered, in the laminar flow settling chamber the gas velocity is parabolic

$$u_x = 2u\left[1 - \left(\frac{2y}{H}\right)^2\right] \tag{5.5}$$

Where　u——the mean velocity across the chamber.

The fractional collection efficiency is written as

$$\eta_1 = \frac{v_t L}{\int_{-H/2}^{H/2} u_x \mathrm{d}y} \tag{5.6}$$

Substituting equation (5.5) into equation (5.6), we find

$$\eta_1 = \frac{3v_t L}{4uH} \tag{5.7}$$

Example 5. 1　Derive the fractional collection efficiency of a cylindrical tube in laminar flow if the velocity distribution is uniform.

Solusion

An element of dr is taken from the section of a cylinder of radius R and length L. The height of this element is $2\sqrt{R^2-r^2}$, as shown in Fig. 5. 2.

The area of the element is

$$\mathrm{d}A = 2\sqrt{R^2 - r^2}\,\mathrm{d}r \tag{5.8}$$

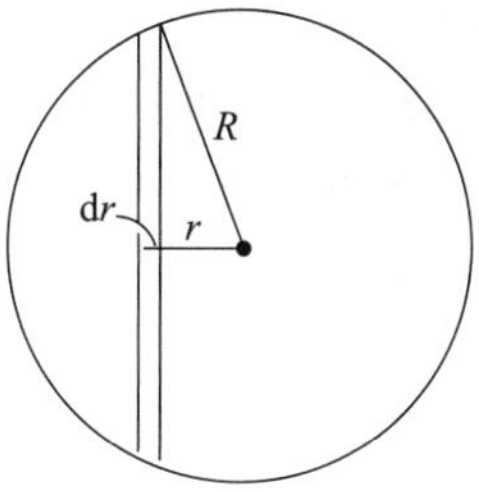

Fig. 5.2 Particle collection of a cylindrical tube in laminar flow

The mass of the particles that are captured in the element is

$$dm = c_0 u \eta_1 dA \tag{5.9}$$

The H in equation (5.4) is substituted by $2\sqrt{R^2 - r^2}$. Then, the collection efficiency of the element is

$$\eta_1 = \frac{v_t L}{2u\sqrt{R^2 - r^2}} \tag{5.10}$$

Substituting equations (5.8) and (5.10) into equation (5.9) and integrating, we obtain the total mass of the collected particles as

$$m = \int_{-R}^{R} c_0 v_t L dr = 2c_0 v_t LR \tag{5.11}$$

The mass get into the cylinder is expressed as

$$M = c_0 u \pi R^2 \tag{5.12}$$

Then, fractional collection efficiency of a cylindrical tube in laminar flow is given by

$$\eta_1 = \frac{m}{M} = \frac{2v_t L}{\pi u R} \tag{5.13}$$

5.1.2 Settling Chamber of Turbulent Flow

The flow in a rectangular channel can be assumed to be turbulent if the Reynolds number Re is greater than 2700. For a duct, the Reynolds number can be defined as

$$\mathrm{Re} = \frac{4R_H \rho u}{\mu} \tag{5.14}$$

Where R_H——the hydraulic radius, defined as the ratio of the cross-sectional area to the perimeter.

Thus, for a duct of height H and width W, the hydraulic radius is written as

$$R_H = \frac{HW}{2(H + W)} \tag{5.15}$$

The gas flow in the air pollution control devices and the ventilation systems in industrial application are often turbulent. The Reynolds number is usually in the order of 10^4 or 10^5. Therefore, it is necessary to discuss the collection efficiency of the turbulent flow settling chamber.

The turbulent flow settling chamber is shown schematically in Fig. 5.3. The derivation is made subject to two assumptions. First, there is a laminar layer adjacent to the bottom surface of the passage into which turbulent eddies do not penetrate, so that any particle which crosses into this layer will be captured shortly. Second, in the remainder of the flow passage the eddying motion due to turbulence will cause a uniform distribution of particle concentration.

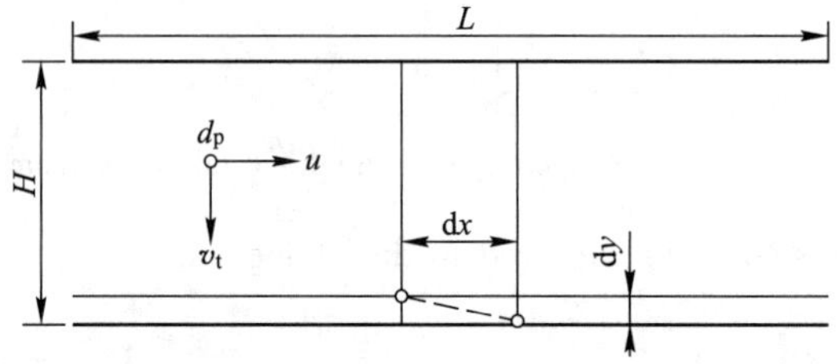

Fig. 5. 3　Separation mechanism of a turbulent flow settling chamber

Consider the element of length dx. If dy represents the thickness of the laminar flow layer, then a particle crosses the dotted line will be surely settle to the bottom while traveling a certain distance downstream in the direction of x. The time dt of the particle traveling forward dx is given by

$$\mathrm{d}t = \mathrm{d}x/u \tag{5.16}$$

During the time dt, the particle settles a distance by

$$\mathrm{d}y = v_t \mathrm{d}t = \frac{v_t \mathrm{d}t}{u} \tag{5.17}$$

When the uniform distribution of all particles across the flow passage produced by the turbulent mixing is assumed, at the entrance to the element dx, the fraction of the particles within the laminar layer dy will equal the ratio of the area inside the laminar layer to the total area. Thus we may write as

$$-\frac{\mathrm{d}c}{c} = \frac{\mathrm{d}y}{H} = \frac{v_t \mathrm{d}x}{uH} \tag{5.18}$$

Integrating equation (5. 18) over the length from the inlet $x = 0$ to the $x = L$, and the concentration from c_i to c, we have

$$\frac{c}{c_i} = \exp\left(-\frac{v_t L}{uH}\right) \tag{5.19}$$

According to the definition of the collection efficiency, we obtain

$$\eta_t = 1 - \frac{c}{c_i} = 1 - \exp\left(-\frac{v_t L}{uH}\right) \tag{5.20}$$

It is noticed that the part of the logarithm of the collection efficiency equation (5. 20) of the turbulent flow settling chamber is the collection efficiency of thelaminar flow settling chamber. That is

$$\eta_t = 1 - \exp(-\eta_l) \tag{5.21}$$

It is not coincident. For example, from the equation(5. 13) we can deduce that the collection efficiency of the cylindrical channel in turbulent flow is

$$\eta_t = 1 - \exp\left(-\frac{2v_t L}{\pi u R}\right) \tag{5.22}$$

Therefore, the method of establishing the collection efficiency model in laminar flow is very useful. Because sometime to derive the collection efficiency of the aerodynamic separation in turbulent flow is very difficult. If the collection efficiency in laminar flow is developed, the collection efficiency of the aerodynamic separation in turbulent flow can be written directly from laminar model.

Example 5.2 Determine the length of a settling chamber required to achieve 80% efficiency for 50μm particles of density $\rho_p = 2 \times 10^3 kg/m^3$ from an air stream of 1m/s at 298K, 1 atm ($1atm = 1.01\times10^5 Pa$). The chamber is to be 1m wide and 1m high.

Solution

We first evaluate the Reynolds number Re for the chamber to determine if the flow will be laminar or turbulent. The dynamic viscosity μ at 298K, 1atm ($1atm = 1.01\times10^5 Pa$) is $1.85\times10^{-5} Pa\cdot s$, and the air density is $1.2kg/m^3$. The hydraulic radius is

$$R_H = \frac{HW}{2(H+W)} = \frac{1}{4}m$$

Then the Reynolds number is

$$Re = \frac{4R_H \rho u}{\mu} = 7.8 \times 10^4$$

Thus the flow is turbulent. The settling velocity of 50μm partide is

$$v_t = \tau g = \frac{\rho_p d_p^2}{18\mu} g = 0.147 m/s$$

From equation (5.20), the length of the settling chamber is

$$L = -\frac{uH}{v_t} \ln(1-\eta) = 10.95m$$

5.2 Inertial Separators

5.2.1 Inertial Deposition in Arch Duct

5.2.1.1 Laminar Flow in Arch Duct

Suppose a flow in an arch duct is laminar, as shown in Fig. 5.4. The gravitational force is neglected in the duct. We shall begin by assuming that the velocity distribution and particle concentration in any cross section are uniform without friction in the flow.

A particle with diameter d_p at the entrance section is located at the radius r_θ. When a particle moves an angle θ with the flow and deposits on the outside wall of the duct at the radius $r = r_2$, as shown as the dotted line in Fig. 5.4, then we say this particle is separated. It is clear that all the particles with diameter d_p at the entrance section of $r > r_\theta$ will be collected. Thus the collection efficiency is

$$\eta_1 = \frac{r_2 - r_\theta}{r_2 - r_1} \tag{5.23}$$

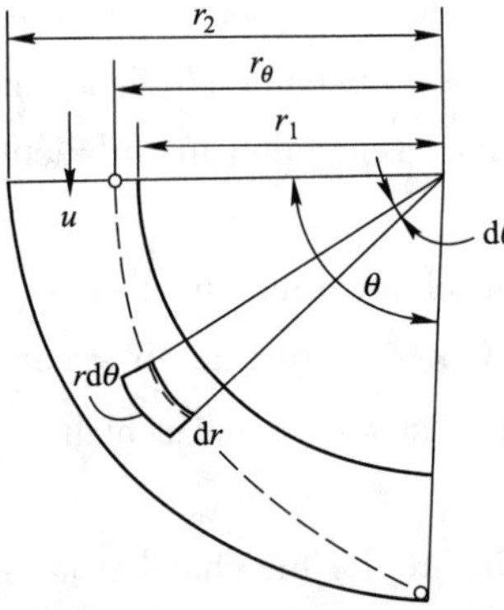

Fig. 5.4 Inertial deposition of particulate in arch duct

If an element of fluid at the point (r, θ) and two-dimensional flow in Fig. 5.4 are considered, the distances of the partide movemens in radial direction and tangential direction in time dt are written as

$$\mathrm{d}r = w\mathrm{d}t, \quad r\mathrm{d}\theta = u\mathrm{d}t \tag{5.24}$$

From equation (5.24), we have

$$\frac{\mathrm{d}r}{\mathrm{d}\theta} = r\frac{w}{u} \tag{5.25}$$

The centrifugal accelerating velocity of the particle, as shown in equation (2.31), is given by

$$w = C_u \tau \frac{u^2}{r}$$

Then equation (5.25) becomes

$$\mathrm{d}r = C_u \tau u \mathrm{d}\theta \tag{5.26}$$

Integrating equation (5.26), it is given by

$$\int_{r_\theta}^{r_2} \mathrm{d}r = \int_0^\theta C_u \tau u \mathrm{d}\theta \tag{5.27}$$

Then we obtain

$$r_2 - r_\theta = C_u \tau u \theta \tag{5.28}$$

Substituting equation (5.28) into equation (5.23), when the velocity distribution is uniform in the arch duct the collection efficiency of inertial deposition of particulate in laminar flow is

$$\eta_1 = \frac{C_u \tau u \theta}{r_2 - r_1} \tag{5.29}$$

If the velocity distribution in the arch duct follows the free vortex, the velocity distribution of unit depth normal to the plane of the paper is given by

$$u = \frac{Q}{r\ln(r_2/r_1)} \tag{5.30}$$

Substituting equation (5.30) into equation (5.27), we find

$$r_2^2 - r_\theta^2 = \frac{2Q}{\ln(r_2 - r_1)} C_u \tau \theta \tag{5.31}$$

Solving out r_θ and substituting r_θ into equation (5.23), we have

$$\eta_l = \frac{1 - \sqrt{1 - 2QC_u\tau\theta/r_2\ln(r_2/r_1)}}{1 - r_1/r_2} \tag{5.32}$$

No matter what kind of gas velocity distribution is, can the fractional collection efficiency of the arch duct in laminar flow be obtained by equations (5.27) and (5.23).

5.2.1.2 Turbulent flow in arch duct

If a flow in the arch duct is turbulent, we first discuss the simpler case of the uniform velocity distribution and particle concentration in the flow. The same method of developing the collection efficiency of the turbulent flow in settling chamber is used. That is, we also assume that there is an laminar layer dr on the outside wall of the duct. All the particles get that into the layer will be captured. Therefore, a relation equation can be presented as

$$\frac{-\,\mathrm{d}c}{c} = \frac{\mathrm{d}r}{r_2 - r_1} \tag{5.33}$$

The thickness of the laminar layer is

$$\mathrm{d}r = w\mathrm{d}t = w\,\frac{r_2 - \mathrm{d}r}{u}\mathrm{d}\theta \approx w\,\frac{r_2}{u}\mathrm{d}\theta = C_c\tau u\mathrm{d}\theta \tag{5.34}$$

Substituting equation (5.34) into equation (5.23), we have

$$\int_{c_i}^{c} \frac{\mathrm{d}c}{c} = -\int_0^\theta \frac{C_c\tau u}{r_2 - r_1}\mathrm{d}\theta \tag{5.35}$$

The solution is given by

$$\frac{c}{c_i} = \exp\left(-\frac{C_c\tau u\theta}{r_2 - r_1}\right) \tag{5.36}$$

According to the definition of the collection efficiency, we obtain

$$\eta_t = 1 - \frac{c}{c_i} = 1 - \exp\left(-\frac{C_c\tau u\theta}{r_2 - r_1}\right) \tag{5.37}$$

We notice that the term of the logarithm $C_u\tau u\theta/(r_2 - r_1)$ in equation (5.37) is the collection efficiency of the arch duct in laminar flow. Actually, this result can be anticipated because the relationship of the collection efficiency in the turbulent flow and the laminar flow has been given by equation (5.21). This regularity is very useful. Since for some cases to derive the efficiency of turbulent flow is very difficult, we can first establish the laminar efficiency equation. And then, the turbulent efficiency can be written directly based on the laminar efficiency. For example, we can write the turbulent efficiency directly from laminar efficiency equation (5.32) of the free vortex gas velocity distribution in the arch duct.

$$\eta_t = 1 - \exp\left(-\frac{1 - \sqrt{1 - 2QC_c\tau\theta/r_2\ln(r_2/r_1)}}{1 - r_1/r_2}\right) \tag{5.38}$$

5.2.2 Cascade Impactor

A common instrument used to measure particle size is the cascade impactor. The cascade impactor typically has several stages. The larger particles are collected on the upper stages, with progressively smaller particles being collected on lower and lower stages[4]. One of impaction stages is shown in Fig. 5.5.

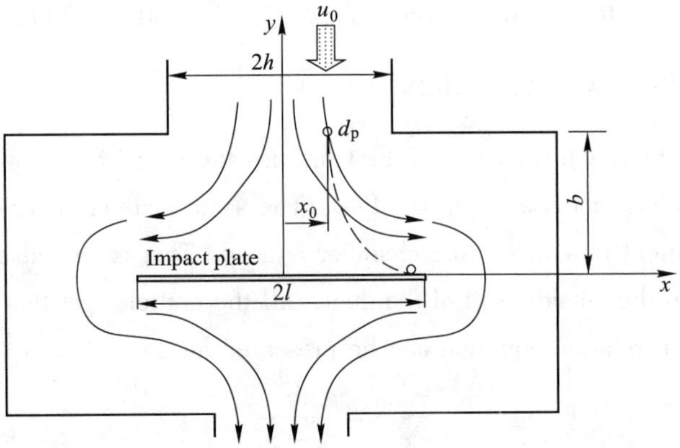

Fig. 5.5 Flow and particle trajectories on the stage of a cascade impactor

The theoretical fractional collection efficiency of the cascade impactor has been not yet developed till now. We have only known that the fractional collection efficiency of the cascade impactor is the function of the Stokes number, wrich is written as

$$S_{tk} = \frac{x_s}{d_s/2} = \frac{x_s}{h} \tag{5.39}$$

Where d_s——the characteristic size.

Here, the characteristic size is the opening of width $2h$ above the impactor plate. x_s is the stopping distance, given by

$$x_s = \tau v_0 \tag{5.40}$$

Where v_0——the initial velocity of the particle.

Substituting equations (5.40) and (2.13) into equation (5.39), we obtain

$$S_{tk} = \frac{\rho_p d_p^2 v_0 C_c}{18\mu h} \tag{5.41}$$

The reason of unknowing the fractional collection efficiency of the cascade impactor to a great extent is that to solve the mathematic model of the fractional collection efficiency under the turbulent condition. Here, we are going to develop a fractional collection efficiency formula of the cascade impactor in turbulent flow.

Since the flow field is symmetrical in Fig. 5.5, we can only consider the effect of the inertial separation in the first quadrant. The flow function is written as

$$\psi(x, y) = -axy \tag{5.42}$$

The gas velocity components are written as

$$u_x = \frac{\partial\psi}{\partial y} = -ax \ , \ u_y = -\frac{\partial\psi}{\partial x} = ay \tag{5.43}$$

Where a——a constant.

Because of $y = b$, $u_y = -u_0$, the constant a can be determined by equation (5.43) as

$$a = -\frac{u_0}{b} \tag{5.44}$$

Then, equation (5.43) becomes

$$u_x = \frac{u_0}{b}x \ , \ u_y = -\frac{u_0}{b}y \tag{5.45}$$

The motion equations of the particle are written as

$$-3\pi\mu d_p(w_x - u_x) = m\frac{dw_x}{dt} \tag{5.46}$$

$$-3\pi\mu d_p(w_y - u_y) = m\frac{dw_y}{dt} \tag{5.47}$$

Where m——the mass of the particle.

w_x, w_y——the velocity components of the particle.

Substituting $m = \pi d_p^3\rho_p/6$ and equation (5.45) into above motion equations, while the relaxation time $\tau = \rho_p d_p^2/18\mu$ is introduced, we have

$$x'' + \frac{1}{\tau}x' - \frac{u_0}{\tau b}x = 0 \tag{5.48}$$

$$y'' + \frac{1}{\tau}y' + \frac{u_0}{\tau b}y = 0 \tag{5.49}$$

The initial conditions of above differential equations are witten as

$$t = 0, \ x = x_0, \ y = b, \ w_x = \frac{dx}{dt} = x' = 0, \ w_y = \frac{dy}{dt} = y' = -u_0 \tag{5.50}$$

The particle trajectory of the particle is obtained by solving the equations (5.48) and (5.49) as

$$x = \frac{1+\alpha}{2\alpha}x_0\exp\left[-\frac{(1-\alpha)t}{2\tau}\right] - \frac{1-\alpha}{2\alpha}x_0\exp\left[-\frac{(1+\alpha)t}{2\tau}\right] \tag{5.51}$$

$$y = \frac{-2\tau u_0 + (\beta+1)b}{2\beta}\exp\left[-\frac{(\beta-1)t}{2\tau}\right] - \frac{2\tau u_0 - (\beta-1)b}{2\beta}\exp\left[-\frac{(\beta+1)t}{2\tau}\right] \tag{5.52}$$

Where

$$\alpha = \sqrt{1 + 4\tau u_0/b} \ , \ \beta = \sqrt{1 - 4\tau u_0/b} \tag{5.53}$$

Because of equation (5.51), $(1+\alpha)\exp\left[-\frac{(1-\alpha)t}{2\tau}\right] \gg (1-\alpha)\exp\left[-\frac{(1+\alpha)t}{2\tau}\right]$, and in equation (5.52), $\frac{-2\tau u_0 + (1+\beta)b}{2\beta}\exp\left[-\frac{(\beta-1)t}{2\tau}\right] \gg \frac{2\tau u_0 - (\beta-1)b}{2\beta}\exp\left[-\frac{(\beta+1)t}{2\tau}\right]$, and the relaxation time τ is very small, $\frac{\tau u_0}{\beta} \to 0$. Then, equations (5.51) and (5.52) can be simplified as

$$x = \frac{1+\alpha}{2\alpha} x_0 \exp\left[-\frac{(1-\alpha)t}{2\tau}\right] \tag{5.54}$$

$$y = \frac{1+\beta}{2\beta} b \exp\left[-\frac{(1-\beta)t}{2\tau}\right] \tag{5.55}$$

If a particle with d_p is collected at the end of the plate $x = l$, as shown as the dotted line, then in the range between the dotted line (the upper limit trajectory of the particle) and $x \geqslant 0$, $y \geqslant 0$, all the particles with d_p are collected. The separation width is x_0. Therefore, the collection efficiency in laminar flow is

$$\eta_l = x_0 / h \tag{5.56}$$

Now, the problem is to find x_0. Suppose t_0 is the time of the particle moving from $y = b$ to $y = 0$ along the dotted line, let $x = l$, $t = t_0$ in equation (5.54), we obtain

$$x_0 = \frac{2\alpha}{1+\alpha} l \exp\left[\frac{(1-\alpha)t_0}{2\tau}\right] \tag{5.57}$$

In equation (5.52), let $y = 0$, and the time t_0 can be found. But it is difficult to get the solution. For conservative calculation, the average moving time of the particles from the opening toward the plate is taken which is written as

$$t_0 \approx b / u_0 \tag{5.58}$$

Combining equations (5.57) and (5.58), the equation becomes

$$\eta_l = \frac{2\alpha}{1+\alpha} \frac{l}{h} \exp\left[\frac{(1-\alpha)b}{2\tau u_0}\right] \tag{5.59}$$

Equation (5.59) is the fractional collection efficiency of the cascade impactor in laminar flow. It is almost impossible to derive the fractional collection efficiency of the cascade impactor in turbulent flow. However, we can use the relation of equation (5.21) to get the fractional collection efficiency of the cascade impactor in turbulent flow easily

$$\eta_t = 1 - \exp\left\{-\frac{2\alpha}{1+\alpha} \frac{l}{h} \exp\left[\frac{(1-\alpha)b}{2\tau u_0}\right]\right\} \tag{5.60}$$

For the cascade impactor as shown in Fig. 5.5, the Stokes number is

$$S_{tk} = \frac{\tau u_0}{h} \tag{5.61}$$

Then the equation (5.61) can also be written as

$$\eta_t = 1 - \exp\left\{-\frac{2\alpha}{1+\alpha} \frac{l}{h} \exp\left[\frac{(1-\alpha)b}{2S_{tk} h}\right]\right\} \tag{5.62}$$

It is proved theoretically that the fractional collection efficiency of the cascade impactor is the function of the Stokes number S_{tk}. The larger S_{tk} is, the higher the collection efficiency.

Example 5.3 The opening width is $2h = 0.02$m, the distance from opening to the plate is $b = 0.1$m, and the length of the impactor plate is $2l = 0.2$m. Calculate the collection efficiency of the cascade impactor in laminar flow and turbulent flow respectively for a particle with 10μm diameter.

Solution

The velocity at the opening is $u_0 = 5\text{m/s}$. The particle density is $\rho_p = 2 \times 10^3 \text{kg/m}^3$.

The relaxation time τ for the particle of 10μm diameter is

$$\tau = \frac{\rho_p d_p^2}{18\mu} = 6 \times 10^{-4} \text{ s}$$

The factor α calculated from equation (5.53) is

$$\alpha = \sqrt{1 + 4\tau u_0 / b} = 1.06$$

According to equation (5.59), the collection efficiency of the cascade impactor in laminar flow is

$$\eta_l = \frac{2\alpha}{1+\alpha} \frac{l}{h} \exp\left[\frac{(1-\alpha)b}{2\tau u_0}\right] = 379\%$$

Because the collection efficiency exceeds 100%, this result in laminar flow is not reasonable. It is shown that the laminar model of the collection efficiency is not suitable for the turbulent flow.

From equation (5.60), the collection efficiency of the cascade impactor for removing the 10μm particle in turbulent flow is

$$\eta_t = 1 - \exp\left\{-\frac{2\alpha}{1+\alpha} \frac{l}{h} \exp\left[\frac{(1-\alpha)b}{2\tau u_0}\right]\right\} = 97.7\%$$

5.3 Cyclone Collector

Cyclone collector (or called cyclone), is gas cleaning device that utilize the centrifugal force created by a spinning gas stream to separate particles from a gas[5]. A standard tangential inlet vertical reverse flow cyclone is shown in Fig. 5.6. The dirty gas flows tangentially into the cyclone at the top. This imparts a swirl flow to the gas as it goes down near the outer radius and then back up in the center core and out the discharge duct at the top. The direction of swirl (clockwise in Fig. 5.6) remains the same in both the outer annular region of flow and the inner core. The particles are slung to the outside, move down the wall, and collected in the dust hopper at the bottom.

5.3.1 Flow Field of Cyclone

Actually, the flow pattern in cyclone is very complex. There are three components: tangential velocity u, axial velocity v, and radical w. The measuring results of the tangential velocity u and axial velocity v had be given by ter Linden, as illustrated in Fig. 5.7.

The tangential velocity u can be described as the quasi free vortex in the outer annular region and quasi forced vortex in the inner core. The interface is located about $r = 2r_1/3$ (r_1 is the radius of the outlet tube). The direction of the axial velocity v is the downward flow in outside region and upward flow in the inner core. The direction of the radical w is the flow from outside to inside.

It is clear that only the particles in the range of the quasi free vortex $r > 2r_1/3$ could possibly be separated. Strauss[6], Leith and Licht[7] thought that there is no separation effect when $r < r_1$.

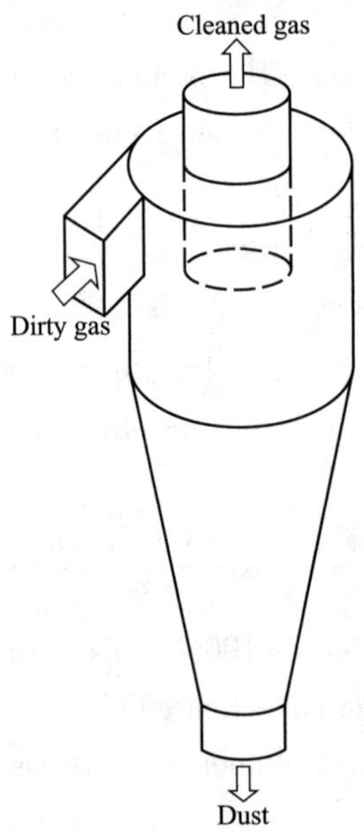

Fig. 5. 6 Isometric view of a tangential inlet reverse-flow cyclone

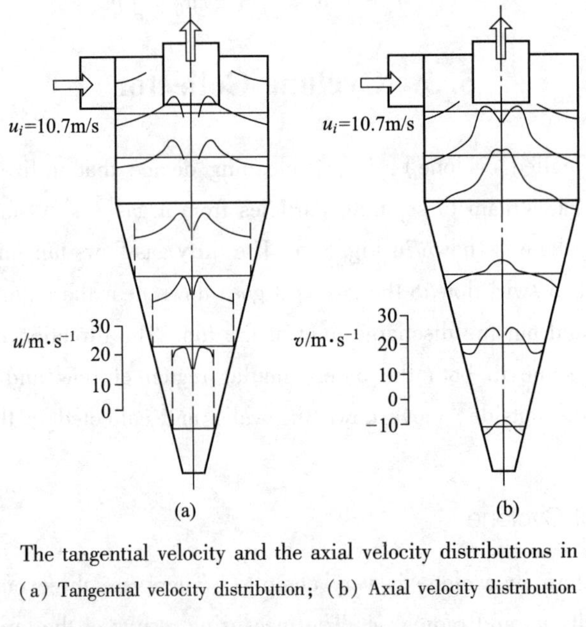

Fig. 5. 7 The tangential velocity and the axial velocity distributions in a cyclone

(a) Tangential velocity distribution; (b) Axial velocity distribution

For the effective separation height, Alexander suggested as

$$L = 7.3r_1\ (r_2^2/ab)^{1/3} \tag{5.63}$$

Where L——the natural length, which is the length from the bottom of the outlet tube to the point of the swirl flow back up;

r_2——the cylindrical body radius;

a, b——the height and width of the inlet duct.

The tangential velocity u is important for airborne particulate separation. In the quasi free vortex region, the tangential velocity can be expressed as

$$u = u_i \left(r_i / r \right)^n \tag{5.64}$$

Where u_i —the inlet gas velocity;

n —a constant, $n = 0.4 \sim 0.9$.

Cheremisinoff et al. thought that it is suitable to choose $n = 0.5$[8], and r_i is the radius of the tangent circle to the axial line of the inlet duct, as shown in Fig. 5. 8, r_i can be calculated by

$$r_i = r_2 - b/2 \tag{5.65}$$

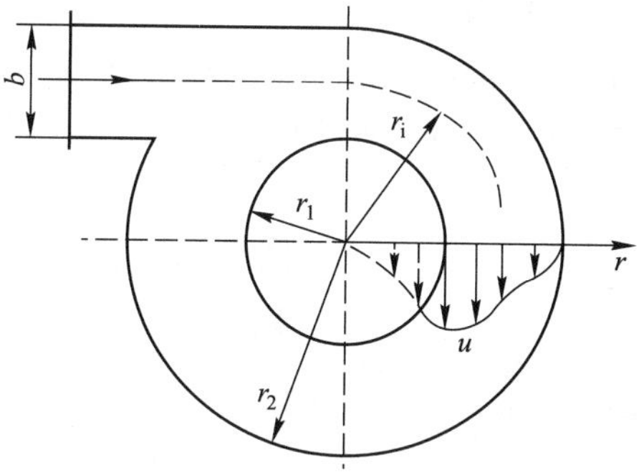

Fig. 5. 8 Definition of the radius r_i and schematic distribution of the tangential velocity

5. 3. 2 Collection Efficiency of Cyclone

Cyclone has been applied in industry for more than one hundred years. The structure of a cyclone is simple. A cyclone is consists of five main parts: inlet duct, cylindrical body, cone, outlet tube, and dust discharging tube. The collection efficiency of cyclone is relatively high. The particles as small as micrometer can be possibly separated by cyclone. Therefore, cyclone is one of the most widely used are pollution control devices today.

The Reynolds number for a cyclone can be defined as

$$\mathrm{Re} = \frac{\rho u}{\mu} \left(\frac{4A_c}{\pi} \right)^{1/2} \tag{5.66}$$

Where A_c —the cross-sectional area of the cyclone body;

$(4A_c/\pi)^{1/2}$ —an equivalent diameter.

From equation (5. 66), it is easy to decide that the flow in cyclone is turbulent. Thus, the turbulent efficiency model for cyclone is much more reasonable.

A great deal of efforts have been devoted to predicting the performance of cyclones theoretically and semi-empirically. Here, we will only present three main turbulent efficiency model: (1) revolution model, (2) boundary layer model, and (3) cut diameter model.

5. 3. 2. 1 Revolution Model

In revolution model, the flow in cyclone is regarded as the turning flow in the arch duct. In the outer annular region $r \geqslant r_1$, assume that the tangential velocity is uniform (although the tangential velocity is not uniform, it is acceptable to use the average velocity $\bar{u}$ for approximate calculation).

Then the equation (5.37) can be applied as

$$\eta = 1 - \exp\left(-\frac{C_u \tau \bar{u} \theta}{r_2 - r_1}\right) \tag{5.67}$$

Where θ——the total turning angle of the vortex flow in cyclone, which was given by Martin as

$$\theta = \frac{2L_1 + L_2}{a}\pi \tag{5.68}$$

Where L_1, L_2——the body length and cone length respectively;

a——the height of the inlet duct.

If the velocity distribution in the outer annular region $r \geqslant r_1$ follows the free vortex flow of equation (5.30), equation (5.38) can be used directly as the collection efficiency of cyclone. That is

$$\eta = 1 - \exp\left(-\frac{1 - \sqrt{1 - 2QC_u \tau \theta / r_2 \ln(r_2/r_1)}}{1 - r_1/r_2}\right) \tag{5.69}$$

5.3.2.2 Boundary Layer Model

When the tangential velocity distribution is described by equation (5.64), a theoretical collection efficiency equation of a cyclone in turbulent flow has been developed Leith and Licht[7]. Because the velocity distribution is quite complicated, Leith and Licht had presented fairly long theoretical derivation of the collection efficiency by means of the turbulent model. Actually, the theoretical result can be derived simply from the laminar model. The analysis process is given as follows.

Substituting equation (5.64) into equation (5.27), neglecting Cunningham slip correction factor C_c, we have

$$\int_{r_\theta}^{r_2} r^n \mathrm{d}r = \int_0^{\theta} \tau u_i r_i^n \mathrm{d}\theta \tag{5.70}$$

The solution is expressed as

$$r_\theta = [r_2^{n+1} - (n+1)\tau u_i r_i^n \theta]^{1/(n+1)} \tag{5.71}$$

Therefore, the collection efficiency of the laminar flow is given by

$$\eta = \frac{r_2 - r_\theta}{r_2 - r_1} = \frac{r_2 - [r_2^{n+1} - (n+1)\tau u_i r_i^n \theta]^{1/(n+1)}}{r_2 - r_1} \tag{5.72}$$

Then, the collection efficiency of the turbulent flow can be fined directly from that of the laminar flow

$$\eta = 1 - \exp\left\{-\frac{r_2 - [r_2^{n+1} - (n+1)\tau u_i r_i^n \theta]^{1/(n+1)}}{r_2 - r_1}\right\} \tag{5.73}$$

This is the collection efficiency of the boundary layer model developed by Leith[7,9]. Where, experimental observations indicate that in a cyclone n ranges between 0.5 and 0.9, depends on the size of the unit and temperature. It has been found experimentally that the exponent n may be estimated from[9]

$$n = 1 - (1 - 0.67D_2^{0.14})\left(\frac{T}{283}\right)^{0.3} \tag{5.74}$$

Where D_2——the cyclone diameter;

T——the gas temperature in Kelvin.

Equation (5.73) has been also simplified by

$$\eta = 1 - \exp(-Md_p^N) \tag{5.75}$$

where,

$$N = 1/(n+1) \tag{5.76}$$

and

$$M = 2\left[\frac{KQ\rho_p(n+1)}{18\mu D_2^3}\right]^{N/2} \tag{5.77}$$

Where K——a geometric configuration parameter that depends only on the relative dimensions of the unit. K=402.9 has been suggested for a tangential inlet reverse-flow cyclone.

5.3.2.3 Cut Diameter Model

The cut diameter model is a more simplified model for collection efficiency of a cyclone. It is assumed that there is a cylindrical surface beneath the outlet duct. As the dust laden gas is spinning, if the centrifugal separation velocityof a particle to the outward is equal to the radical gas velocity inward at the cylindrical surface, the possibility of this particle being captured is 50%. That is, the collection efficiency of this size particle is 50%. Thus, at the cylindrical separation surface we have

$$w_p + w = 0 \tag{5.78}$$

If Cunningham slip correction factor C_c is neglected, the centrifugal separation velocity of the particle w_p, as shown in equation (2.31), is given by

$$w_p = \tau\frac{u^2}{r_c} = \frac{\rho_p d_c^2}{18\mu}\frac{u^2}{r_c} \tag{5.79}$$

Where d_c——cut diameter, the diameter of the particle which could be captured with 50% possibility;

r_c——the radius of assumed cylindrical surface. Barth had taken the radius of assumed cylindrical surface r_c as the radius of the outlet duct r_1.

The average radical gas velocity w is

$$w = \frac{Q}{2\pi r_1 L} \tag{5.80}$$

Where L——the natural length, given by equation (5.63).

Then, equation (5.78) becomes

$$\frac{\rho_p d_c^2}{18\mu}\frac{u^2}{r_1} - \frac{Q}{2\pi r_1 L} = 0 \tag{5.81}$$

And the calculation equation for cut diameter is expressed as

$$d_c = \sqrt{\frac{9\mu Q}{\pi\rho_p L u^2}} \tag{5.82}$$

According to the turbulent theory, the collection efficiency follows

$$\eta = 1 - \exp(-kd_p^m) \tag{5.83}$$

Where k ——a constant.

Because the collection efficiency is 50% (when $d_p = d_c$), we have

$$1 - \exp(-kd_c^m) = 50\% \tag{5.84}$$

Then the solution of k is given by

$$k = \frac{0.693}{d_c^m} \tag{5.85}$$

Thus the collection efficiency for the cut diameter model is given by

$$\eta = 1 - \exp[-0.693\,(d_p/d_c)^m] \tag{5.86}$$

Actually, equation (5.86) is a semi-empirical expression for the collection efficiency of cyclone. The exponent m is in the range between 1 and 2. If the exponent m is suggested to be 1 in the cut diameter model[10], then equation (5.86) becomes

$$\eta = 1 - \exp(-0.693d_p/d_c) \tag{5.87}$$

Obviously, it is not reasonable that m is selected to be 1 because from the theoretical analysis result in an arch duct, as shown in equation (5.37), the collection efficiency of cyclone is the exponential function of the relaxation time τ. While the τ is proportional to the square of the particle diameter. Thus, it would be more strict theoretically that the exponent m is 2 rather than 1. Therefore, equation (5.86) becomes

$$\eta = 1 - \exp[-0.693\,(d_p/d_c)^2] \tag{5.88}$$

5.3.3 Pressure Drop of Cyclone

A major factor in the evaluation of cyclone design and performance is the pressure drop undergone by the gas in traversing the cyclone. The power which must be expended somewhere in the dust removal system to overcome this pressure drop, is given by

$$W = Q\Delta p \tag{5.89}$$

The prediction of pressure drop in a cyclone is at best only an approximation to the actual value. Thus, a performance test of the actual cyclone is required to determine the pressure accurately. However, the accurate theoretical pressure drop calculation equation has not yet found till now.

An empirical expression given by Louis Theodore, as inaccurate as it may be, will be chosen for presentation as

$$\Delta p = 14.75\frac{Q^2}{D_1^2ab\,(L_1L_2/D_2^2)\,1/3} \tag{5.90}$$

Where Q——the flow rate;

D_1, D_2——the outlet duct diameter and cyclone body diameter respectively;

a, b——the height and width of the inlet duct;

L_1, L_2——the body length and cone length respectively.

Another more rough approximate prediction for the cyclone pressure drop is given by

$$\Delta p = \xi \frac{\rho u_i^2}{2} \tag{5.91}$$

Where u_i——the velocity in the inlet duct (usually in the range between 15m/s and 25m/s);

ξ——a coefficient which ranges between 6 and 9.

For a standard cyclone with long cone as shown in Fig. 5.6, the coefficient of ξ can be taken as 9 for conservative design. The pressure drop is in the range from 1000~1500Pa.

5.3.4 Dimensions of Cyclone

Cyclone collection efficiency increases with increasing particle size, particle density, inlet gas velocity, and cyclone body length. On the other hand, cyclone collection efficiency decreases with increasing cyclone diameter, gas outlet duct diameter, and gas inlet area. For any specific cyclone whose ratio of dimensions is fixed, the collection efficiency increases as the cyclone diameter decreased. The design of a cyclone represents a compromise among collection efficiency, pressure drop, and size. Higher efficiency requires higher pressure drop and longer body.

The dimensions of a tangential-entry, reverse-flow cyclone exist optinal ratio. The dimensions are shown in Fig. 5.9.

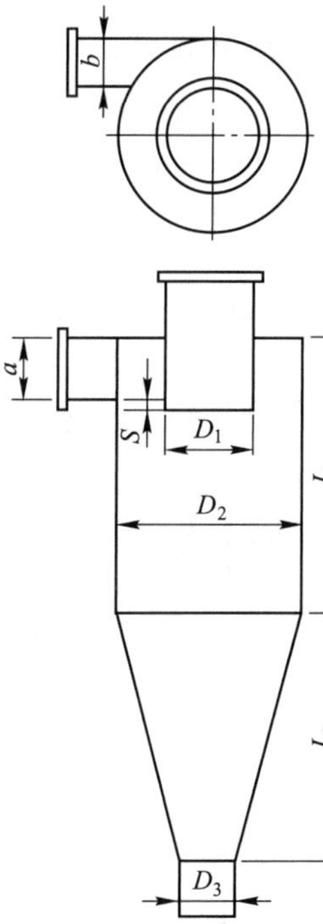

Fig. 5.9 Geometric specification for the design of a cyclone

Shepherd and Lapple determined 'optimum' dimensions for cyclone[11]. All dimensions were related to the body diameter D_2. A common set of specifications is listed in Table 5.1. Because these dimension ratios are just a reference in cyclone design, of course, the dimensions of a cyclone is not necessarily designed exactly as the same as the ratios listed in Table 5.1.

Table 5.1 Specifications of the dimensions for tangential inlet reverse-flow cyclone design

Dimension	a	b	D_1	L_1	L_2	D_3	S
Ratio	$D_2/2$	$D_2/4$	$D_2/2$	$2D_2$	$2D_2$	$D_2/4$	$D_2/8$

5.3.5 Multiple Cyclone

Efficiency of a cyclone increases as the diameter of the cyclone is reduced even if the tangential velocity remains constant. To achieve higher efficiencies dictates the use of smaller cyclone. But if the cyclone is made smaller while the tangential velocity remains about the same, the flow rate which the cyclone can handle is reduced by the square of the cyclone diameter. Very small cyclones have been used to collect small particles from very small gas flows for research and gas sampling purposes. However, the industrial problem is to treat large gas flows. Several practical schemes have been worked out to place a large number (up to several thousands) small cyclones in parallel[12], so that they can treat a large gas flow and capturing smaller particles. The most common of these arrangements, called a multiple cyclone, is sketched in Fig. 5.10.

Many small cyclones in the multiple cyclone are mass-produced and inserted into sheet metal supporters. In Fig. 5.10, the circular gas motion in each cyclone is caused by a set of sheet metal turning vanes that replace the tangential inlet duct of an ordinary cyclone. The gas outlet tubes are connected to a common gas outlet header. The problem in the multiple cyclone is that the flow rate of the dust-laden gas is needed to be distributed equally to every small cyclone.

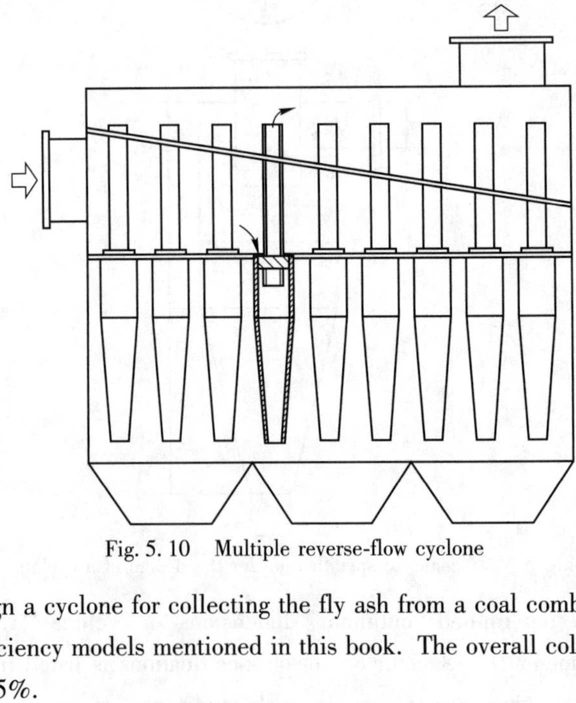

Fig. 5.10 Multiple reverse-flow cyclone

Example 5.4 Design a cyclone for collecting the fly ash from a coal combustion boiler according to three turbulent efficiency models mentioned in this book. The overall collection efficiency of the cyclone required is 85%.

The original design data are the flow rate of the gas is $5000 m^3/h$, the gas temperature is 180℃, the density of fly ash is $2000 kg/m^3$, and the particle size distribution is given in Table 5.2.

Table 5.2 Particle size distribution of the fly ash discharged from a coal combustion boiler

Size range $\Delta d_p/\mu m$	1~5	5~10	10~30	30~60	60~80	>80
Average size $d_p/\mu m$	3	7.5	15	45	70	90
Fraction of mass $\Delta F/\%$	6	12	22	29	18	13

Solution

The steps of design are given as following:

(1) Determine the sizes of the inlet duct. The inlet velocity of a cyclone usually ranges between 15~25m/s. If the inlet velocity u_i = 18m/s is chosen as a trial value, the inlet duct area can be found as

$$A_i = \frac{Q}{3600u_i} = 0.077\text{m}^2$$

According to Table 5.1, $a=D_2/2$, $b=D_2/4$, thus we know that a is equal to $2b$. Because $A_i = ab$, then the height a and width b are

$$a = 0.38\text{m}, \quad b = 0.19\text{m}$$

The real velocity in the inlet duct is

$$u_i = \frac{Q}{3600ab} = 19.2\ \text{m/s}$$

(2) Determine other dimensions of the cyclone. According to Table 5.1, we find

$$D_2 = 0.76\text{m}, \quad D_1 = 0.38\text{m}, \quad S = 0.1\text{m}, \quad D_3 = 0.19\text{m}, \quad L_1 = L_2 = 0.76\text{m}$$

(3) Calculate the fractional collection efficiency and overall collection efficiency.

1) *D* Revolution model. Based on Revolution model, the fractional collection efficiency from equation (5.67) is (Cunningham slip correction factor C_c is neglected because the particle size is larger than 1μm)

$$\eta_R = 1 - \exp\left(-\frac{\tau u_i \theta}{r_2 - r_1}\right)$$

where, r_2 = 0.38m, r_1 = 0.19m.

The total turning angle of the vortex flow in cyclone is

$$\theta = \frac{2L_1 + L_2}{a}\pi = 6\pi$$

The relaxation time is

$$\tau = \frac{\rho_p d_p^2}{18\mu}$$

When the gas temperature is 180℃, the dynamic viscosity μ is 2.5×10^{-5} Pa · s. The calculated values of the relaxation time is filled in Table 5.3. The fractional collection efficiency can be written as

$$\eta_R = 1 - e^{-1904\tau}$$

The calculated values of the fractional collection efficiency is also filled in Table 5.3. The overall collection efficiency is

$$\eta_{RT} = \sum \Delta F_i \eta_{Ri} = 87.2\%$$

2) Boundary layer model. Based on the boundary layer model, the fractional collection efficiency from equation (5.75) is

$$\eta_B = 1 - \exp[-Md_p^{1/(n+1)}]$$

where, the exponent n is determined by equation (5.74) as

$$n = 1 - (1 - 0.67D_2^{0.14})\left(\frac{T}{283}\right)^{0.3} = 0.591 \approx 0.6$$

From equation (5.77), we obtain

$$M = 2\left[\frac{KQ\rho_p(n+1)}{18\mu D_2^3}\right]^{1/2(n+1)}$$

where, the geometric configuration parameter K=402.9.

Then

$$M = 2\left[\frac{KQ\rho_p(n+1)}{18\mu D_2^3}\right]^{1/2(n+1)} = 2586$$

The fractional collection efficiency of the boundary layer model in this problem is given by

$$\eta_B = 1 - \exp(-2586 d_p^{0.625})$$

The calculated values of the fractional collection efficiency is given in Table 5.3. And the overall collection efficiency is

$$\eta_{BT} = \sum \Delta F_i \eta_{Bi} = 93.2\%$$

3) Cut diameter model. Based on the cut diameter model, the fractional collection efficiency from equation (5.88) is

$$\eta = 1 - \exp[-0.693(d_p/d_c)^2]$$

From equation (5.63), the natural length is

$$L = 7.3r_1(r_2^2/ab)^{1/3} = 1.75\text{m}$$

And from equation (5.82), the cut diameter is

$$d_c = \sqrt{\frac{9\mu Q}{\pi\rho_p L u^2}} = 8.8\mu\text{m}$$

The values of the fractional collection efficiencies calculated by the cut diameter model is given in Table 5.3 either. The overall collection efficiency is

$$\eta_{CT} = \sum \Delta F_i \eta_{Ci} = 84.3\%$$

Table 5.3 Calculation of the fractional efficiency in the cyclone design

Size range $\Delta d_p/\mu m$	1~5	5~10	10~30	30~60	60~80	>80
Average size $d_p/\mu m$	3	7.5	15	45	70	90
Fraction of mass $\Delta F/\%$	6	12	22	29	18	13
Relaxation time $\tau/\times10^{-4}$s	0.4	2.5	17.8	45.0	109.0	180.0
Fractional efficiency η_R	0.073	0.401	0.966	0.999	1.000	1.000
Fractional efficiency η_B	0.599	0.802	0.918	0.993	0.998	0.979
Fractional efficiency η_C	0.077	0.396	0.866	0.999	1.000	1.000

4) Pressure drop prediction. When the gas temperature is 180℃, the gas density is 0.8kg/m^3. From equation (5.91), we have

$$\Delta p = \xi \frac{\rho u_i^2}{2} = 9 \times \frac{0.8 \times 19.2^2}{2} = 1327\ (\text{Pa})$$

The final results of the cyclone design are listed in Table 5.4.

Table 5.4 Results of the cyclone design for separating the fly ash from a boiler

a/m	b/m	D_1/m	D_2/m	L_1/m	L_2/m	D_3/m	S/m	u_i/m · s^{-1}	η_T/%	Δp/Pa
0.38	0.19	0.38	0.76	0.76	0.76	0.19	0.10	19.2	≥85	1327

From the results obtained from the three turbulent efficiency models, we can see that only the cut diameter model is a little bet lower than 85%. It will be acceptable in cyclone design.

Comparing above three collection efficiency models from this example, we have also found that the overall collection efficiency calculated by the boundary layer model is much higher than the others. The result predicted by the boundary layer model could be possibly overestimated. Thus, it would be better to use the revolution model or the cut diameter model to get the approximate value of the collection efficiency in the practical cyclone design.

Exercises

5.1 A sodium hydroxide in air at standard condition is to be collected in a gravity settling chamber. The unit is 4m wide, 3m high, and 6m long. And the inlet gas velocity is 1m/s. Calculate the smallest mist droplet (spherical in shape) that will be entirely collected by this settler where the laminar flow model is assumed. The density of the mist droplets is assumed to be equal to 1.2×10^{3}kg/m^3.

5.2 A gravity settling chamber is 50cm wide×100cm long with 20 plates, the channel thickness is 0.3cm, and the flow rate is 1.2m^3/h. It is observed that it operates at an efficiency of 70% for collecting 5μm. How many plates would be required to have the unit operate at 80% efficiency?

5.3 Installation of a gravity settling chamber to remove limestone particles (density = 2.67×10^3 kg/m^3) from an air standard condition has been proposed. The inlet dust loading is 2g/m^3, and the flow rate 100m^3/h. The inlet size distribution of the limestone is shown in Table 5.5.

Table 5.5 Mass fraction of the inlet limestone particles

Particle size range/μm	Mass fraction/%	Particle size range/μm	Mass fraction/%
0~5	2	50~100	28
5~20	6	100~500	36
20~50	17	≥500	11

The dimensions of the unit are 0.6m wide, 0.5m high, and 2.0m long. Calculate an average collection efficiency over each size range.

5.4 A house shape gravity settling chamber is shown in Fig. 5.11. Try to establish the efficiency calculation equation of laminar and turbulent flow respectively.

5.5 In Fig. 5.4, the gas velocity distribution in an arch channel is $u = 2Q/r$. Try to establish the efficiency

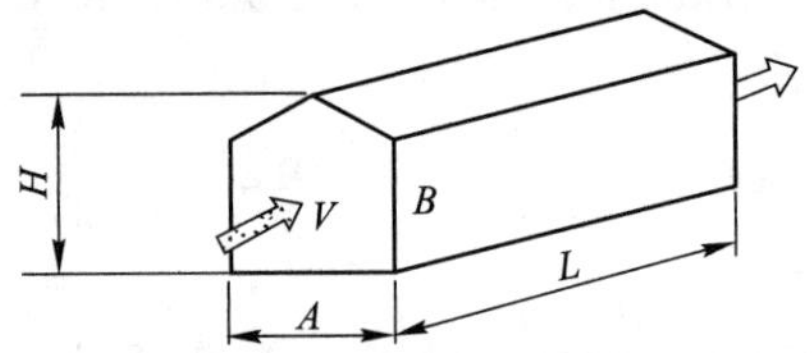

Fig. 5. 11　A gravity settling chamber in house shape

calculation equation of turbulent flow at angle θ.

5. 6 In example 5. 3, if 99% efficiency of a cascade impactor for collecting the particles of 10μm is 99%, what is the opening of width $2h$ needed? If the efficiency of a cascade impactor for collecting the particles of 5μm is 90%, what is the opening of width $2h$ needed? The particle density $2.5\times10^3\,kg/m^3$ and turbulent flow in the cascade impactor are assumed.

5. 7 If a certain cyclone has a cut size of 9μm, what will be the overall collection efficiency? Plot the fractional efficiency curve of this cyclone if the particle size distribution of dust from a cement plant is shown in Table 5. 6.

Table 5. 6　Mass fraction of the cement particles

Particle size range/μm	Mass fraction/%	Particle size range/μm	Mass fraction/%
0. 0~2. 5	15	20. 0~30. 0	17
2. 5~5. 0	8	30. 0~40. 0	14
5. 0~10. 0	12	>40. 0	21
10. 0~20. 0	15		

5. 8 Calculate the collection efficiency of a cyclone on treating 10μm particle according to revolution model and cut diameter model. The calculation data are shown in Table 5. 7.

Table 5. 7　Original data for collection efficiency calculation mass

Parameters for calculation	Values	Parameters for calculation	Values
Gas density	$1.2\,kg/m^3$	Inlet duct height	0. 2m
Gas viscosity	$1.8\times10^{-5}\,Pa\cdot s$	Outlet tube radius	0. 1m
Particle density	$2.0\times10^3\,kg/m^3$	Cylinder height	0. 4m
Inlet gas velocity	20. 0m/s	Cylinder radius	0. 2m
Gas flow rate	$0.4\,m^3/s$	Cone height	0. 6 m
Inlet duct width	0. 1m		

5. 9 A cyclone is used to treat a gas stream containing $5.2\,g/m^3$ of dust. If $1.2\,g/m^3$ of dust escapes, what is the overall collection efficiency? Plot the fractional efficiency curve when the following data are shown in Table 5. 8.

Table 5.8 Inlet and outlet particle size distribution of a cyclone

Particle size range/μm	Size distribution analysis (mass fraction)/%	
	Inlet	Outlet
0~5	6.2	21.3
5~10	9.4	29.9
10~20	13.8	31.6
20~50	22.9	15.1
>50	47.7	2.1

5.10 A cyclone is used to clean the pollutant stream which contains particles of two sizes, such as large and small. The fractional collection efficiencies are shown in Table 5.9. Please find the overall collection efficiency. If it is connected another cyclone with the same performance in series, how about the overall collection efficiency of the system?

Table 5.9 Mass fraction of inlet particles in polluted gas

Particle size	Mass fraction/%	Fractional efficiencies/%
Large	70	90
Small	30	30

References

[1] Jaworek A, Krupa A, Czech T, et al. Modern electrostatic devices and methods for exhaust gas cleaning: A brief review [J]. Journal of Electrostatics, 2007, 65 (3): 133-155.

[2] Ragland K W, Han J, Aerts D J, et al. Performance of an aerodynamic particle separator [J]. Waste Management, 1996, 16 (8): 735-740.

[3] Isai A M, Widodo W A. Studi numerik karakterisasi aliran 3 dimensi multifase (gas-solid) pada gravity settling chamber dengan variasi kecepatan inlet dan diameter partikel pada aliran dilute phase [J]. Jurnal Teknik ITS, 2013, 2 (2).

[4] Mercer T T, Tillery M I, Newton G J, et al. A multi-stage, low flow rate cascade impactor [J]. Journal of Aerosol Science, 1970, 1 (1): 9-15.

[5] Leith D, Mehta D. Cyclone performance and design [J]. Atmospheric Environment, 1973, 7 (5): 527-549.

[6] Strauss, W. *Industrial Gas Cleaning. Second Ed* [M]. New York, Pergamon Press. 1975.

[7] Leith D, Licht W. The collection efficiency of cyclone type particle collectors-A new theoretical approach [J]. ALCHE Symposium Ser., 1972, 68 (126): 196-206.

[8] Chermisinoff P N, Young R A. *Air Pollution Control and Design Handbook* [M]. New York, Marcel Dekker Inc., 1977.

[9] Licht W. *Air Pollution Control Engineering : Basic Calculations for Particulate Collection* [M]. New York, Marcel Dekker, Inc., 1988.

[10] Buonicore, J, Davis W T. *Air Pollution Engineering Manual* [M]. New York, Van Nostrand Reinhold, 1992.

[11] Cooper, C D, Alley F C. *Air Pollution Control: A design Approach* [M]. Boston, PWS Engineering, 1986.

[12] Noh S, Heo J, Woo S, et al. Performance improvement of a cyclone separator using multiple subsidiary cyclones [J]. Powder Technology, 2018, 338: 145-152.

6 Electrostatic Precipitation

6.1 Basic Principles of ESP

If the aerodynamic force (inertial or centrifugal force) could not separate the particles below 5μm, other more powerful than inertial forces, such as the electric force, can be utilized to work on the smaller particles. A common device of particulate separation based on the electric force is called electrostatic precipitator (ESP)[1].

Electrostatic precipitator is one of the most widely used collection devices for particulates. It has many advantages. Its range of size is enormous. It is used in large fossil fuel fired power plants and small household air-conditioning system as well. It is versatile enough to provide virtually complete collection of particles of many substances, both solids and liquids. Its power requirements are low, and its efficiency is very high[2]. For these reasons, the electrostatic precipitator is the preferred collection method where high efficiency is required with small particles.

6.1.1 Types of ESP

The basic idea of all ESPs is to give the particles an electrostatic charge and then put them in an electric field that drives them to a collecting wall. In one type of ESP, called a two-stage ESP[3], charging and collecting are carried out in separate parts of the ESP, as shown in Fig. 6.1. This type is often used in building air conditioners. However, for most industrial applications the two steps are carried out simultaneously in the same part of the ESP which is called single-stage ESP[4], as shown in Fig. 6.2.

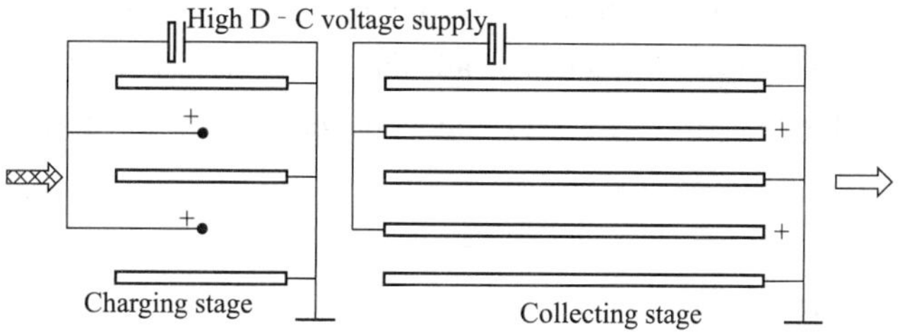

Fig. 6.1 Two-stage ESP

The metal plates are used for the collection electrodes in Fig. 6.1 and Fig. 6.2. They also called wire-plate ESP. If the collection electrode of an ESP is cylindrical, it is called the wire-tube ESP[5], as shown in Fig. 6.3. The wire-tube ESPs are often used for clean the wet smoke nowadays in industrial application.

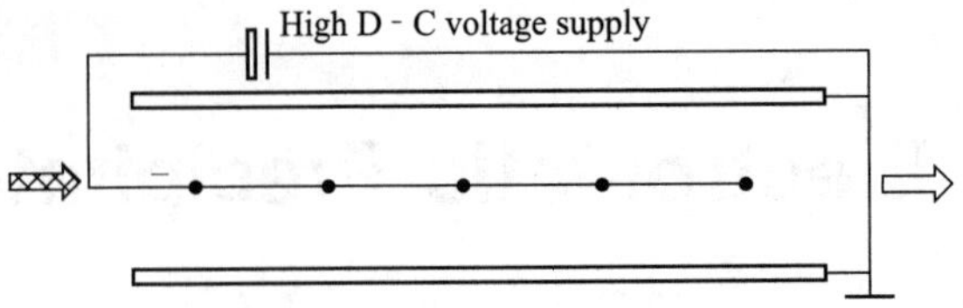

Fig. 6. 2　Single-stage ESP

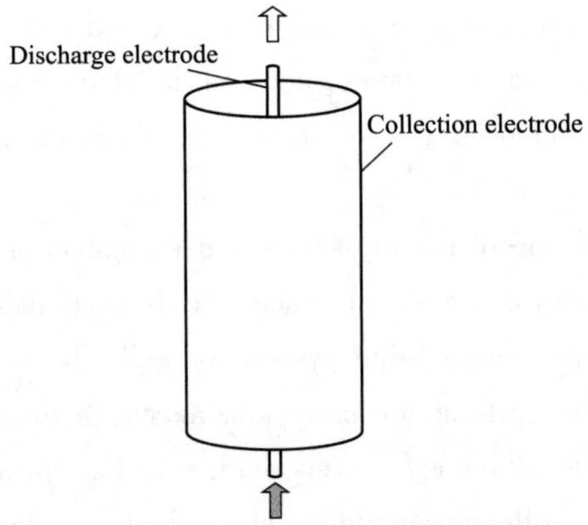

Fig. 6. 3　Single-stage wire-tube ESP

6. 1. 2　Particulate Collection Process of ESP

When the high D-C voltage is supplied between the wires and plate electrodes, the electric field is set up in Fig. 6. 2. When the particles get into this electric field, they are charged by means of a corona surrounding a highly charged wire electrode. An electric force acted on the charged particles is formed in the electric field. The electric force causes the particles to migrate to the collection electrodes where they stick. The collected particles are removed from the plate electrodes by shaking (rapping) mechanially. The removed particles then go down along the collection electrodes to the hopper located beneath the collection section.

In the electric field between the wire and collection electrode, the force acted on a charged particle based on the electrostatics is given by

$$F_E = qE_p \tag{6.1}$$

Where F_E——the electric force;

q——the charge on the particle;

E_p——the collection electric field strength.

Equation (6. 1) is the fundamental principle of ESP. It is clear that the charge q on the particle and the collection electric field strength E_p are required to be determined.

6. 2 Particle Charging

6. 2. 1 Corona and Ion Generation in ESP

There are several mechanisms of formation of electrons or ions for particle charging. However, the only mechanism found practicable for use in an ESP is the corona[6]. If a corona wire is applied negative voltage, in the outside corona region, negative ions are produced. If a corona wire is applied positive voltage, in the outside corona region, positive ions produced. The phenomenon of the negative corona discharging is described as follows.

It is fortunate that there is a few of ions in theatmosphere. These ions are formed from the air atoms by incident background radiation which knocks an electron off from the atom. Thus, a few free electrons and an equal number of positive ions will be present in the air at all times.

Suppose if a negative high voltage is supplied on the wire electrode, and an electric field is set up, an electron will be accelerated in the direction toward the grounded electrode. If this electron has sufficient energy, upon collision with the molecule, it may knock an electron off from the air molecule. That is to say, producing a second electron which will be also accelerated by electric field, as well as leaving a positive ion behind. Now, there are two electrons. Two electrons again knock two extra electron off from another two molecules. At this point, four electrons are present. Then, four electrons become eight, and so on. This electron producing process continues very rapidly. At last, many thousands of free electrons are produced, as shown in Fig. 6. 4. This electron producing process is referred to as electron avalanche, which is accompanied by emission of light and sometime by generation of sound.

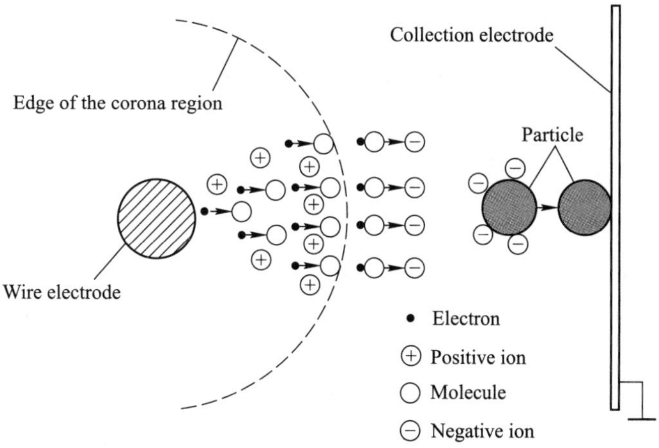

Fig. 6. 4 Corona generating and particle charging

This process occurs within a narrow region surrounding the wire electrode. Under suitable condition, a spark discharging will occur around the wire electrode. The radius at the outer edge of the corona will not be treated rigorously since its determination requires knowledge of the field strength variation across the corona region. Cobine had indicated that the following empirical equation may be used to estimate the radius of the corona region r_a[7], written as

$$r_a = r_0 + 0.03\sqrt{r_0} \tag{6.2}$$

Where r_0——the radius of the wire electrode.

When the electrons move into a region of much lower electric filed strength, they will be slowed down by collisions and will be unable to knock the electrons off from the molecules. Molecules of a number of gases have the property that they absorb electrons which collide with them especially if the electrons have relatively low energies upon impact. Thus, negative ions are formed, as shown in Fig. 6.4, and the number of electrons is reduced. Under typical condition of negative corona operation, the electrons are almost entirely eliminated and a large number of negative ions is formed.

6.2.2 Charge on a Particle

When the negative ions move in the electric field toward the collecting electrode and encounter particles to which they become attached. If the negative ions come in contact with the particles, it is called negatively charged particles. Thus, the charged particles in the field will move toward the collecting electrode and finally will be collected by the collecting electrode.

Once an electric field and current density are established, particle charging can take place. Particle charging is essential to the precipitation process because a particle to migrate toward the collection electrode is directly proportional to the charge on the particle. The most significant factors influencing particle charging are particle diameter, applied electric field, current density, and exposure time.

The particle charging process can be attributed mainly to two physical mechanism, field charging and diffusion charging, or called thermal charging. The classical equations of calculating the charge on a particle by field charging and diffusion charging are respectively[7]

$$q_f = 3\pi\varepsilon_0 E_q d_p^2 \left(\frac{\varepsilon}{\varepsilon + 2}\right)\left(\frac{1}{1 + \tau_q/t}\right) \tag{6.3}$$

$$q_d = \frac{2\pi\varepsilon_0 k_B T d_p}{e}\ln\left(1 + \frac{d_p \rho_e k E_q}{8\varepsilon_0 k_B T}t\right) \tag{6.4}$$

Where E_q——the charging electric field streng thin V/m;

ε_0——the permittivity of free space, $\varepsilon_0 = 8.85 \times 10^{-12}$ C/(V · m);

ε——the relative permittivity of particle, for non-metal solid material, $\varepsilon = 5 \sim 7$;

τ_q——time constant in s;

T——gas gas temperature in K;

t——the exposure time of particle in the electric field in s;

k_B——Boltzmann constant in J/K;

k——the ion mobility in $m^2/(V \cdot s)$;

ρ_e——the charge density in gas in m^{-3};

e——the charge of a electron, $e = 1.6 \times 10^{-19}$C.

Thus, the total charge on a particle is

$$q = q_f + q_d \tag{6.5}$$

For the field charging of equation (6.3), time constant $\tau_q = (4\varepsilon_0/N_0ek) \ll t$, it means that the charge on a particle can get to the saturation charge very quickly, then equation (6.3) becomes

$$q_f = 3\pi\varepsilon_0 E_q d_p^2 \left(\frac{\varepsilon}{\varepsilon + 2}\right) \tag{6.6}$$

For the diffusion charging of equation (6.4), the charge density is hard to be determined because it is changing cross the electric field. Considering the effect of the mean free path of gas, Cochet[3] had proposed a equation of charge on a particle of combination of field charging and diffusion charging

$$q = \pi\varepsilon_0 E_q d_p^2 \left[\left(\frac{\varepsilon - 1}{\varepsilon + 2}\right)\left(\frac{2}{1 + 2\lambda/d_p}\right) + (1 + 2\lambda/d_p)^2\right] \tag{6.7}$$

When $d_p \gg \lambda$, equation (6.7) becomes equation (6.6). Generally, if $d_p > 1\mu m$, it is well approximate to calculate the particle charge by equation (6.6). If $d_p \leqslant 1\mu m$, the charge of a particle can be calculated by equation (6.7).

Example 6.1 Calculate the saturation charge and number of the electrons on a particle with radius of 1μm for the following data.

$$E_q = 500\text{kV/m},\ \varepsilon_0 = 8.85 \times 10^{-12}\text{C/(V·m)},\ \varepsilon = 6,\ e = 1.6 \times 10^{-19}\text{C}$$

Solution

From equation (6.6), the saturation charge is

$$q = 3\pi\varepsilon_0 E_q d_p^2 \left(\frac{\varepsilon}{\varepsilon + 2}\right) = 3.13 \times 10^{-17}\ \text{C}$$

The number of the electrons is

$$n = \frac{q}{e} = 196$$

6.3 Electric Field

According to the shape of the collection electrode, the ESPs can be classified into two types, such as cylindrical ESP and parallel plate ESP. The electric field of the industrial ESPs is often not uniform. Since the charge of a particle and the electric force act on the particle are the function of electric field strength, it is necessary to discuss the distribution of electric field between the corona electrodes and collecting electrodes[8].

6.3.1 Electric Field in Wire-tube ESP

A cylindrical ESP, sometimes called wire-tube ESP, is illustrated in Fig. 6.5. When a high voltage is applied between the wire and the tube electrodes, the electric field is set up. A current will flow across the electric field because the electrons, ions, and charges on the particles are present in the electric field. The electric field with space charges is described by Possion

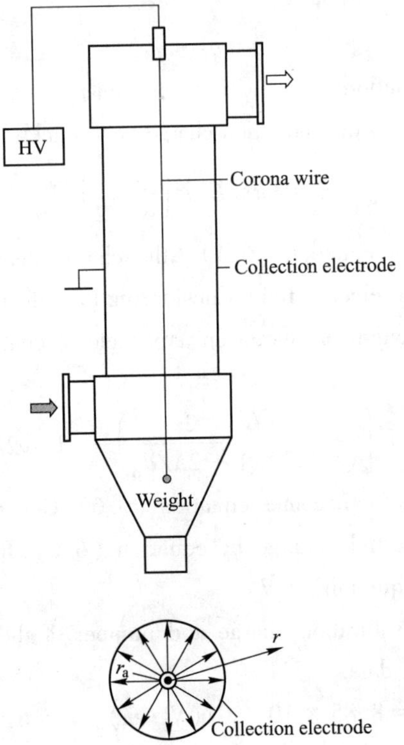

Fig. 6. 5　Cylindrical ESP

equation as

$$\nabla^2 E = \frac{\rho_e}{\varepsilon_0} \tag{6. 8}$$

Under the axisymmetric condition in radical coordination, equation (6. 8) becomes

$$\frac{\mathrm{d}(rE)}{r\mathrm{d}r} = \frac{\rho_e}{\varepsilon_0} \tag{6. 9}$$

The current per square meter j cross the electric field is written as

$$j = \rho_e v_e \tag{6. 10}$$

Where　j——the current density in $\mathrm{A/m^2}$;

v_e——the migration velocity of ion, given by

$$v_e = kE \tag{6. 11}$$

Where　k——ion mobility which depends on the temperature and pressure in electric field, given by

$$k = k_0 \sqrt{\frac{T}{T_0}\left(\frac{1 + S/T_0}{1 + S/T}\right)\frac{p_0}{p}} \tag{6. 12}$$

In equation (6. 12), k_0 is the ion mobility at standard condition (283K, 1atm). In dry air, for the negative ion, $k_0 = 2.10\mathrm{cm^2/(V \cdot s)}$, and for positive ion, $k_0 = 1.32\mathrm{cm^2/(V \cdot s)}$. S is Surtherland constant. In dry air, $S = 330$. T_0 and p_0 are the temperature and pressure at standard condition, and T and p are the temperature and pressure of the real gas.

From equations (6.10) and (6.11), the charge density in the electric field is

$$\rho_e = \frac{j}{kE} \tag{6.13}$$

The relation of the current per square meter j and the current per meter of wire i is

$$j = \frac{i}{2\pi r} \tag{6.14}$$

Where i——the current density in A/m.

Then the charge density in the electric field can be written as

$$\rho_e = \frac{i}{2\pi rkE} \tag{6.15}$$

Substituting equation (6.15) to equation (6.9), we have

$$\frac{r}{2}\frac{\mathrm{d}E^2}{\mathrm{d}r} + E^2 - \frac{i}{2\pi\varepsilon_0 k} = 0 \tag{6.16}$$

The solution of equation (6.16) is

$$E = \sqrt{\frac{C^2}{r^2} + \frac{i}{2\pi\varepsilon_0 k}} \tag{6.17}$$

Where C——a constant.

Suppose the field strength is E_0 at the edge of the corona r_a. Then, the constant C can be determined by equation (6.17) as

$$C = r_a\sqrt{E_0^2 - \frac{i}{2\pi\varepsilon_0 k}} \tag{6.18}$$

where, r_a is given by equation (6.2).

Then equation (6.17) becomes

$$E = \sqrt{\frac{r_a^2 E_0^2}{r^2} + \frac{i}{2\pi\varepsilon_0 k}\left(1 - \frac{r_a^2}{r^2}\right)} \tag{6.19}$$

The field strength E_0 of triggering the corona at the edge of the corona $r = r_0$ is given empirically by Peek[9] as

$$E_0 = 3 \times 10^6 f(\delta + 0.03\sqrt{\delta / r_a}) \tag{6.20}$$

$$\delta = T_0\, p / T\, p_0 \tag{6.21}$$

Where, f is a roughness factor that accounts for rough spaces on the wire surface. The effect of roughness is to reduce the field strength need to form the corona. For the clean smooth wire, $f = 1$, for practical application, $f = 0.6 \sim 0.7$ is reasonable value to use in the absence of other information. under the standard conditions ($T_0 = 293\mathrm{K}$, $p_0 = 1\mathrm{atm}$) because of

$$E = -\frac{\mathrm{d}U}{\mathrm{d}r} \tag{6.22}$$

Substituting equation (6.22) into equation (6.19) and integrating equation (6.19), the electric potential difference is

$$U = r_a E_0 \left\{ \ln \frac{r_c}{r_a} - 1 - \left[1 + \frac{i}{2\pi\varepsilon_0 k} \left(\frac{r_c}{r_a E_0} \right)^2 \right]^{1/2} + \ln \left\{ 1 + \left[1 + \frac{i}{2\pi\varepsilon_0 k} \left(\frac{r_c}{r_a E_0} \right)^2 \right]^{1/2} \right\} /2 \right\} \tag{6.23}$$

where, r_c is the cylindrical electrode radius. If let $i = 0$, the lowest value of the voltage U_0 applied between the wire electrode and cylindrical collection electrode to trigger the corona is

$$U_0 = r_a E_0 \ln(r_c / r_a) \tag{6.24}$$

In industrial ESP, the current per length of the wire i is very small, the second term of equation (6.23) can be unfolded into series. The first term of the series is selected, that is

$$\ln \frac{1}{2} \left\{ 1 + \left[1 + \frac{i}{2\pi\varepsilon_0 k} \left(\frac{r_c}{r_a E_0} \right)^2 \right]^{1/2} \right\} \approx \frac{1}{2} \left[1 + \frac{i}{2\pi\varepsilon_0 k} \left(\frac{r_c}{r_a E} \right)^2 \right]^{1/2} \tag{6.25}$$

Then from equations (6.23), (6.24) and (6.25), the current per length of the wire i is found by

$$i = \frac{8\pi\varepsilon_0 k}{r_c^2 \ln(r_c / r_a)} U(U - U_0) \tag{6.26}$$

Then, the field strength E can be determined from equation (6.19) according to the current per length of the wire i calculated by equation (6.26) if applied voltage of U has been known.

6.3.2 Electric Field in Wire-plate ESP

Fig. 6.6 gives a qualitative representation of the electric field distribution and electric potential surfaces in a wire-plate geometry which is commonly used. Although the electric field is very nonuniform near the wire, it becomes essentially uniform near the collection plates. The current density is also very nonuniform throughout the electrode space and the maximum along a line from the wire to the plate.

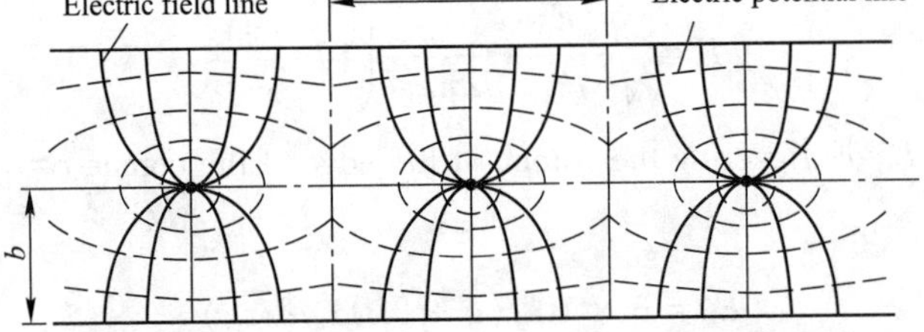

Fig. 6.6 Electric field configuration for wire-plate ESP

Many researchers have discussed the electric field distribution in the space between the wire and plate electrodes. Comperman had developed a theoretical solution of the electric field distribution in wire-plate ESP which is a series of arc hyperbolic function. It is quite difficult to calculate the electric field strength mathematically by Compermans' theory[9].

Here, we assume the electric field strength distribution follows Gaussian model. Firstly, we derive the electric field strength distribution in a pair of wire-plate electrodes. And then, the electric field strength distribution between multi-wire electrode and a plate electrode can be developed based on the superposition principle.

When there is only one wire electrode in wire-plate electrodes, the distribution of the electric lines is schematically shown in Fig. 6.7. The nature of the electric line distribution shows that the electric field strength reduces with increasing x. On the surface of the wire, the electric field strength reach the maximum. The electric field strength reduces will reduce very quickly with increasing y. When $y \rightarrow \infty$, the electric field strength $E \rightarrow 0$. The shape of the electric field strength distribution looks like Fig. 6.8.

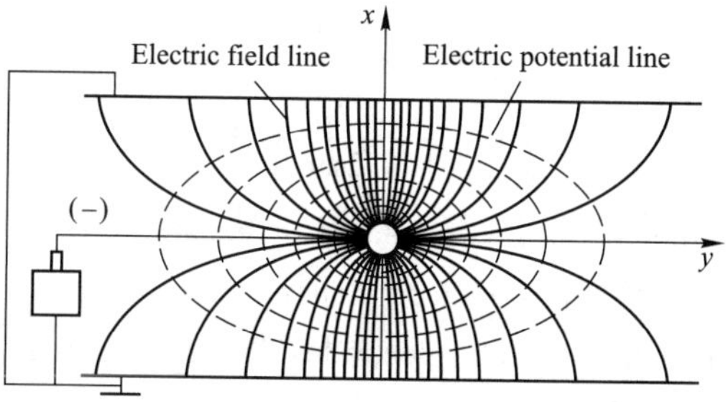

Fig. 6.7 Distributions of the electric field line and potential line in a pair of wire-plate electrodes

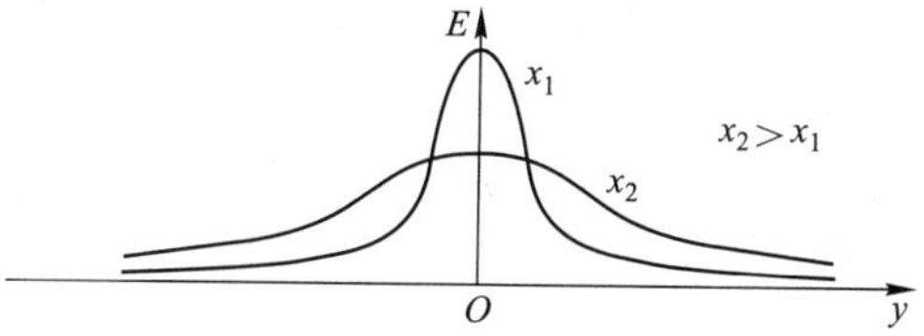

Fig. 6.8 Electric field strength distribution in a pair of wire-plate electrodes

From the shape of the field strength distribution, we can assume that electric field strength distribution E should follow the Gaussian distribution function (normal distribution) which is described by

$$E(x, y) = \frac{A}{\sqrt{2\pi}\sigma}\exp\left(-\frac{y^2}{2\sigma^2}\right) \tag{6.27}$$

Where A——a constant which is determined by boundary condition;

σ——the deviation which is the function of x.

According to the Gaussian flux theorem, as shown in Fig. 6.7, the flux on the edge surface of the corona region r_a must be equal to the flux on two arbitrary infinite planes $\pm x$. When the symmetry is considered, we have

$$\oint_s E_0 ds = 2\int_{-\infty}^{\infty} E(x, y) dy \tag{6.28}$$

where, E_0 is determined by Peek equation (6.20).

Substituting equation (6.27) into equation (6.28), we yield

$$A = \pi r_a E_0 \tag{6.29}$$

Then, equation (6.27) can be written as

$$E(x, y) = \frac{\pi r_a E_0}{\sqrt{2\pi}\sigma}\exp\left(-\frac{y^2}{2\sigma^2}\right) \tag{6.30}$$

Now we determine the deviation σ. The variation tendency of the electric field strength with x along the x axis ($y = 0$) is almost the same as the electric field strength of wire-tube electrode. Then the radius r in equation (2. 19) is replaced by x, we have

$$E(x, 0) = \sqrt{\frac{r_a^2 E_0^2}{x^2} + \frac{i}{2\pi\varepsilon_0 k}\left(1 - \frac{r_a^2}{x^2}\right)} \quad r_a \leqslant x \leqslant b \tag{6.31}$$

Substituting equation (6. 31) into equation (6. 30), we find

$$\sigma = \frac{\pi r_a E_0}{\sqrt{2\pi\left[\frac{r_a^2 E_0^2}{x^2} + \frac{i}{2\pi\varepsilon_0 k}\left(1 - \frac{r_a^2}{x^2}\right)\right]}} \tag{6.32}$$

Then the electric field strength can be calculated by equation (6. 30) when the deviation σ is determined by equation (6. 32). It is noticed that equation (6. 30) satisfy the Gaussian flux theorem at any plane parallel to the plate electrodes. It means that the electric field distribution which is assumed to be a normal distribution is reasonable.

For the wire-plate ESP consisting of a series of parallel wires stretched between two parallel plates, as shown in Fig. 6. 6, according to the superposition principle, the electric field distribution in the wire-plate ESP is given by

$$E(x, y) = \frac{\pi r_a E_0}{\sqrt{2\pi}\sigma}\left\{\sum_{m=1}^{m} \exp\left\{-\frac{[y - 2(m-1)c]^2}{2\sigma^2}\right\} + \sum_{n=1}^{n} \exp\left[-\frac{(y + 2nc)^2}{2\sigma^2}\right]\right\} \tag{6.33}$$

Where m——the number of the central wire and the wires on the right of the central wire;

n——the number of the wires on the left of the central wire.

The electric field distribution in the wire-plate ESP is shown schematically in Fig. 6. 9.

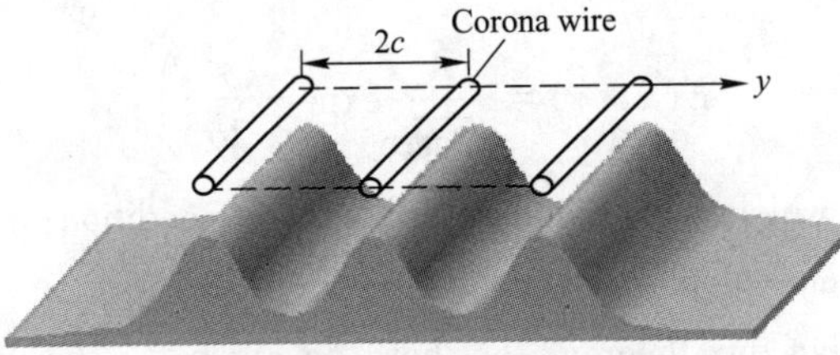

Fig. 6. 9 Outlook of the electric field distribution in the wire-plate ESP

If there is only one wire, which is $m=1$, $n=0$, equation (6. 33) becomes equation (6. 30). It is noticed that the electric field strength E reduces very quickly with distance y increasing. That is, when $y \geqslant 2c$, $E \to 0$. Therefore, only the superposition effect of adjacent two wires is considered. Then, when many wires are located between two collection electrodes, equation (6. 33) can be simplified approximately as

$$E(x, y) = \frac{\pi r_a E_0}{\sqrt{2\pi}\sigma}\left\{\exp\left[-\frac{y^2}{2\sigma^2}\right] + \exp\left[-\frac{(y - 2c)^2}{2\sigma^2}\right] + \exp\left[-\frac{(y + 2c)^2}{2\sigma^2}\right]\right\} \quad (-c \leqslant y \leqslant c) \tag{6.34}$$

Thus equation (6. 34) can be used to predict the electric field strength at any point (x, y) of the wire-plate ESP.

Fig. 6. 10 (a) was the numerical simulation results of the electric field distribution in a wire-plate ESP given by Oglesby et al. [9] Under the same condition, the values of the electric field distribution in a wire-plate ESP calculated by the new equation (6. 34) is shown in Fig. 6. 10 (b). The comparison shows that both results have a good agreement.

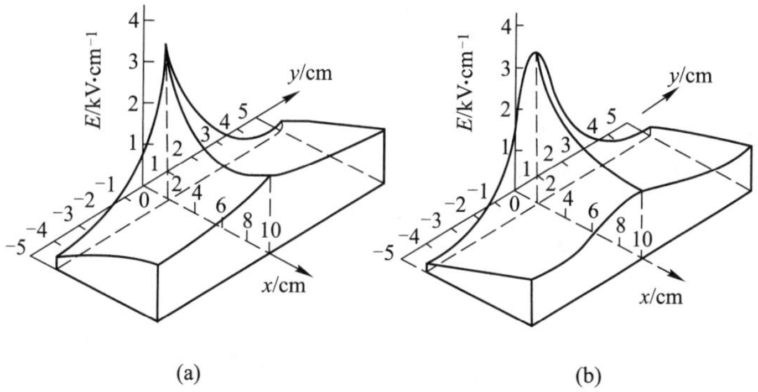

Fig. 6. 10 Comparison of electric field distributions in a wire-plate ESP
(a) Numerical simulation given by Oglesby; (b) Results calculated by equation (6. 34)

Theoretically, if the charge on a particle q and the electric field strength E have been determined from above discussion, the electrostatic force F_E can be predicted by equation (6. 1).

Example 6. 2 The configuration of a wire-plate ESP is shown in Fig. 6. 11. The distance between wire and plate is 0. 15m. At the standard condition, when the applied voltage is 60kV, the measured current density is 0.1×10^{-3} A/m, and the radius of the clean smooth wire is 1mm. Calculate the electric field strength at $y = 0$m, 0. 05m, 0. 075m, and 0. 15m on the plane of $x = 0.1$m. Draw a curve of the electric field distribution on the plane of $x = 0.1$m. [$\varepsilon_0 = 8.85 \times 10^{-12}$ C/(V · m), $k = 2.1 \times 10^{-4}$ m²/(V · s)].

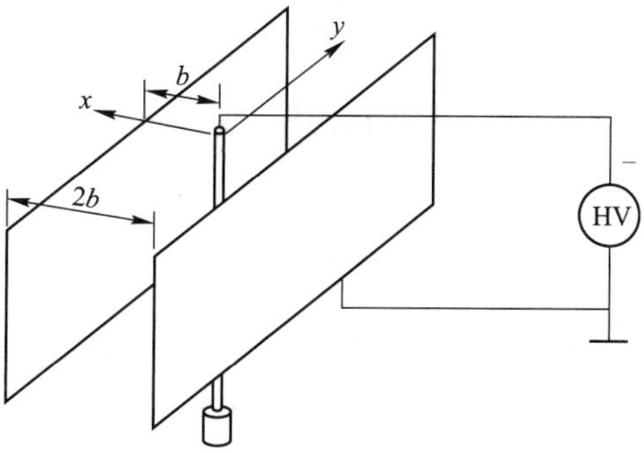

Fig. 6. 11 Configuration of a wire-plate ESP with one corona wire

Solution

Find the radius of the corona region r_a. From equation (6. 2), the radius of the corona region is

$$r_a = r_0 + 0.03\sqrt{r_0} = 2 \times 10^{-3} \text{ m}$$

Find the field strength E_0 at the edge of the corona. From equation (6.20), the field strength E_0 is

$$E_0 = 3 \times 10^6 f(\delta + 0.03\sqrt{\delta/r_a}) = 5 \times 10^6 \text{ V/m}$$

Determine the deviation σ on the plane of $x=0.1$m. From equation (6.32), we have

$$\sigma = \frac{\pi r_a E_0}{\sqrt{2\pi\left[\frac{r_a^2 E_0^2}{x^2} + \frac{i}{2\pi\varepsilon_0 k}\left(1 - \frac{r_a^2}{x^2}\right)\right]}} = 0.092$$

Calculate the electric field strength at $y=0$m, 0.05m, 0.075m and 0.15m on the plane of $x=0.1$m. From equation (6.32), we obtain

$$E(x, y) = \frac{\pi r_a E_0}{\sqrt{2\pi}\sigma}\exp\left(-\frac{y^2}{2\sigma^2}\right)$$

Then, we find

$$E(0.1, 0) = 1.36 \times 10^5 \text{ V/m}$$
$$E(0.1, 0.05) = 1.17 \times 10^5 \text{ V/m}$$
$$E(0.1, 0.075) = 0.98 \times 10^5 \text{ V/m}$$
$$E(0.1, 0.15) = 0.36 \times 10^5 \text{ V/m}$$

The electric field distribution on the plane of $x=0.1$m is shown in Fig. 6.12.

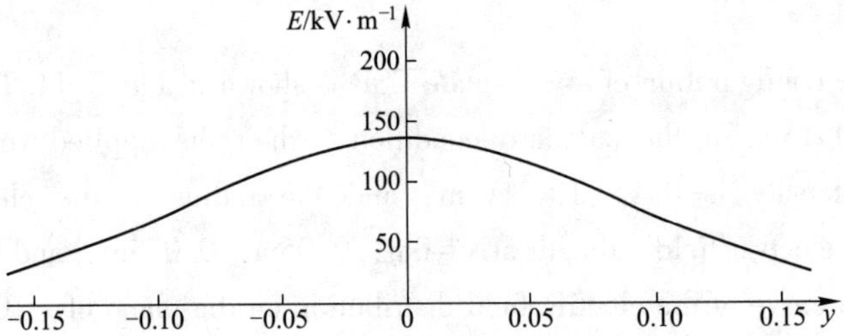

Fig. 6.12　Electric field distribution on the plane of $x=0.1$m

($U=60$kV, $i=0.1\times10^{-3}$A/m, $b=0.15$m)

6.4 Deutsch Equation

6.4.1 Collection Efficiency of Wire-plate ESP

It has been known that in ESP the flow is the turbulent flow. So that, as before, the particle concentration is uniform at any point across the device due to the turbulent flow. we will suppose the existence of a thin layer adjacent to the collection electrodes across which the particle migration collection occur. The only migration occurs across a layer to the collection electrode, because the turbulent mixing in the core of the flow overwhelms the tendency of particles to migrate. Thus, the turbulent flow ESP is quite analogous to the turbulent flow settling chamber, only the physical mechanism leading to particle migration differs.

Assume that the charge q on a particle and the collection electric field strength E_p are known. The electrical migration velocity of a particle diameter d_p is given by equation (2.34) as

$$\omega = E_p q C_c / 3\pi\mu d_p \tag{6.35}$$

Where E_p——collection electric field strength.

The process to develop the collection efficiency equation of an ESP is the same as that of the settling chamber of equation (5.20). The collection efficiency is given by

$$\eta = 1 - \exp\left(-\frac{L\omega}{bv}\right) \tag{6.36}$$

Where L——the length of the collection electrode;

b——the distance between the corona wire and collection electrode;

v——the gas velocity in the electric field.

Assume that H is the height of the collection electrode. Then collection electrode area is $A = LH$, and the gas volume flow rate is $Q = bHv$. Thus, equation (6.36) can also be written as

$$\eta = 1 - \exp\left(-\frac{A}{Q}\omega\right) \tag{6.37}$$

Equation (6.36), or (6.37) is called Deutsch equation. Equation (6.37) was first used in an empirical form in 1919 by Anderson and derived theoretically by Deutsch in 1922 (White[10], 1963). Therefore, it has generally been referred to as the Deutsch equation and sometimes as the Deutsch-Anderson equation.

6.4.2 Collection Efficiency of Wire-tube ESP

Though the collection efficiency equation (6.36), or (6.37), of the wire-plate ESP can be developed by the same method of establishing the collection efficiency of settling chamber, someone may suspect that the collection efficiency equation of the wire-tube ESP could also be expressed by equation (6.37). White had prove that the collection efficiency of the wire-tube ESP is also equation (6.37) based on the probability which seems to be not easily understood. Here we use the turbulent flow model to derive the collection efficiency of the wire-tube ESP.

We consider a section of the wire-tube ESP with length dx, as shown in Fig. 6.13. According to the assumption of the turbulent flow model, there is a layer dr adjacent to the cylindrical electrode. The migration of the charged particles only occurs across a layer to the collection electrode. The reduced increment of the concentration $-$dc is equal to the fraction of the particles collected in the region dr. We have

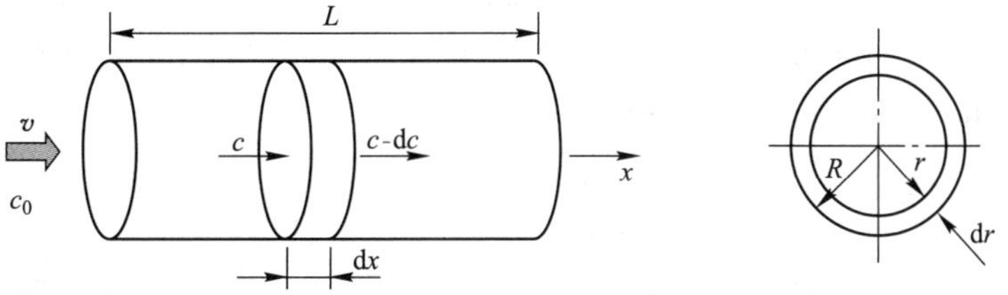

Fig. 6.13 Theoretical analysis of the collection efficiency for wire-tube ESP

$$\frac{-\mathrm{d}c}{c}=\frac{2\pi(R-\mathrm{d}r)\mathrm{d}r}{\pi R^2} \tag{6.38}$$

Where c——the particle concentration at the entrance of the section $\mathrm{d}x$.

Since $\mathrm{d}r=\omega\mathrm{d}t=\omega\mathrm{d}x/v$, and $\pi R^2 v=Q$, when the high level minim is neglected on the right side of equation (6.38), equation (6.38) can be written as

$$\frac{-\mathrm{d}c}{c}=\frac{2\pi R\omega\mathrm{d}x}{Q} \tag{6.39}$$

Integrating equation (6.39) for c from c_0 to c, and for x from 0 to L, we have

$$\frac{c}{c_0}=\exp\left(-\frac{2\pi R\omega L}{Q}\right)=\exp\left(-\frac{A\omega}{Q}\right) \tag{6.40}$$

Where A——the collection area of a cylindrical tube.

According to the definition of the collection efficiency, we obtain

$$\eta=1-\frac{c}{c_0}=1-\exp\left(-\frac{A\omega}{Q}\right) \tag{6.41}$$

Thus, it has been proved that the collection efficiency equations of the wire-plate ESP and the wire-tube ESP in the turbulent flow are identical.

6.5 Effect Factors on Collection Performance of ESP

Although the Deutsch equation can be used to estimate the collection efficiency of an ESP, the assumption of constant ω is overly restrictive. It must be pointed out. However, even though it is possible to derive theoretically the electric field and migration velocity in an ESP with well-defined geometry, the idealized conditions corresponding to the theory seldom exist in actual practice. Factors such as particle re-entrainment and gas channeling around the collecting zones cannot be accounted for theoretically[11-13]. Because of these uncertainties, industrial ESP design is often based on empirical migration velocities for using the Deutsch equation. Nevertheless, it is still importantto understanding fundamental relationships among the variations in an ESP.

6.5.1 Charging Electric Field Strength

The migration velocity of a charged particle is the function of q in equation (6.35). From equation (6.6) or (6.7), the charge on a particle depends on the electric field strength E_q.

In wire-tube ESP as shown in Fig. 6.5, the field strength is symmetrically distributed. It is reasonable to take the average electric field strength between the wire and tube wall as the charging electric field strength. Since $r \gg r_a$. From equation (6.19), the average charging electric field strength in the wire-tube ESP is expressed as

$$E_q=\frac{1}{R}\int_{r_a}^{R}\sqrt{\frac{r_a^2E_0^2}{r^2}+\frac{i}{2\pi\varepsilon_0 k}}\mathrm{d}r \tag{6.42}$$

However, the charging electric field strength of the wire-plate ESP is different from that of the wire-tube ESP. In wire-plate ESP as shown in Fig. 6.7, it is clear that all the particles will going through the maximum E along a plane from the wire to the plate. We assume that the charge on a particle can reach the saturation charge very quickly when the particles go through the plane from the wire to the plate. According to equation (6.34), the average charging electric field strength in the wire-plate ESP is given by

$$E_q = \frac{1}{b}\int_{r_a}^{b} E(x, 0)\,dx = \frac{1}{b}\int_{r_a}^{b} \frac{\pi r_a E_0}{\sqrt{2\pi}\,\sigma}\left[1 + 2\exp\left(-\frac{c^2}{\sigma^2}\right)\right]dx \tag{6.43}$$

Where σ——the function of x, which is determined by (6.32).

To solve the equation (6.43), a numerical integration has to be used. For an acceptable approximate calculation, the average charging electric field strength in the wire-plate ESP can be calculated by

$$E_q = \frac{U}{b} \tag{6.44}$$

6.5.2 Collection Electric Field Strength

Due to the turbulent flow assumption, it is assumed that the migration of the charged particles only occurs across a layer near the collection electrode. Therefore, the collection electric field strength refers to the electric field strength near the collection electrode.

For the wire-plate ESP, because the field strength is not uniform, we have to take the average value of the electric field strength E on the surface of the plate electrode ($x=b$) as the collection electric field strength E_p. According to equation (6.34), the collection electric field strength is

$$\begin{aligned} E_p &= \frac{1}{c}\int_0^c E(b, y)\,dy \\ &= \frac{\pi r_a E_0}{c\sqrt{2\pi}\,\sigma}\int_0^c \left\{\exp\left[-\frac{y^2}{2\sigma^2}\right] + \exp\left[-\frac{(y-2c)^2}{2\sigma^2}\right] + \right. \\ &\quad \left. \exp\left[-\frac{(y+2c)^2}{2\sigma^2}\right]\right\} dy \end{aligned} \tag{6.45}$$

Notice that $b \gg r_a$, from equation (6.32), the deviation σ is

$$\sigma = \frac{\pi r_a E_0}{\sqrt{2\pi\left(\frac{r_a^2 E_0^2}{b^2} + \frac{i}{2\pi\varepsilon_0 k}\right)}} \tag{6.46}$$

It is still quite difficult to calculate the collection electric field strength in wire-plate ESP from equation (6.45). Due to the superposition of the electric field caused by many wires in the wire-plate ESP as shown in Fig. 6.6, the electric field strength near the collection electrode becomes essentially uniform. The electric field strength $E(b, 0)$ can be represented the collection electric field strength approximately. From equation (6.34), the collection electric field strength is expressed as

$$E_p = E(b, 0) = \frac{\pi r_a E_0}{\sqrt{2\pi}\,\sigma}\left[1 + 2\exp\left(-\frac{c^2}{\sigma^2}\right)\right] \tag{6.47}$$

Example 6.3 Estimate the charging electric field strength and the collection electric field strength in the wire-plate ESP at the standard conditions. The data for calculation are given as following.

$U = 40\text{kV}$, $b = 0.1\text{m}$, $c = 0.1\text{m}$, $r_0 = 1 \times 10^{-3}$ m, $\varepsilon_0 = 8.85 \times 10^{-12}$ C/(V · m), $k = 2.1 \times 10^{-4}\text{m}^2/(\text{V} \cdot \text{s})$.

Solution

From equation (6.44), the average charging electric field strength is

$$E_q = \frac{U}{b} = 400\text{kV/m}$$

From equation (6.2), the radius of the corona region is

$$r_a = r_0 + 0.03\sqrt{r_0} = 2 \times 10^{-3}\text{m}$$

From equation (6.20), the critical field strength E_0 for triggering the corona is

$$E_0 = 3 \times 10^6 f(\delta + 0.03\sqrt{\delta/r_a}) = 5 \times 10^6\text{V/m}$$

From equation (6.24), the voltage U_0 of triggering the corona is

$$U_0 = r_a E_0 \ln(b/r_a) = 39.1\text{kV}$$

Then from equation (6.26), the current per length of the wire i is found

$$i = \frac{8\pi\varepsilon_0 k}{b^2 \ln(b/r_a)} U(U - U_0) = 0.043 \times 10^{-5}\text{ A/m}$$

From equation (6.46), the deviation σ is

$$\sigma = \frac{\pi r_a E_0}{\sqrt{2\pi\left(\frac{r_a^2 E_0^2}{b^2} + \frac{i}{2\pi\varepsilon_0 k}\right)}} = 0.125$$

Form equation (6.47), the collection electric field strength is

$$E_p = E(b,\ 0) = \frac{\pi r_a E_0}{\sqrt{2\pi}\sigma}\left[1 + 2\exp\left(-\frac{c^2}{\sigma^2}\right)\right] = 206\text{kV/m}$$

From example 6.3, it is known that the collection electric field strength is much lower than the charging electric field strength.

6.5.3 Dust Re-entrainment

The dust re-entrainment or particle re-suspension in the boundary layer resulting from turbulent fluid flow is important in collection processes of ESPs. In deriving Deutsch equation of the turbulent flow model, it was assumed that there is no dust re-entrainment in the separating layer near the collection electrodes. However, in the operation of a ESP, the collection electrode rapping and the aerodynamic force of gas flow will lead to dust re-entrainment[14]. Here, we just discuss the effect of aerodynamic re-entrainment on particle collection.

The lift force for a sphere with diameter d_p resting on a flat wall in wall-bounded shear and in pressure difference have been already given by equation (2.45) in section 2.3, we rewrite here as

$$L_s = C_s \frac{1}{8} \pi \rho d_p^2 v^2 \tag{6.48}$$

$$L_p = \frac{1}{21} \pi \rho d_p^2 \left(\frac{d_p}{2b} \right)^{2/7} v_{max}^2 \tag{6.49}$$

Where v —the average velocity;

v_{max} —the maximum velocity at the axes of the duct.

For the wire-plate ESP, the relation of average velocity v and the maximum velocity v_{max} is written as

$$v = \frac{7}{8} v_{max} \tag{6.50}$$

For the wire-tube ESP, the relation of average velocity v and the maximum velocity v_{max} is written as

$$v = \frac{19}{21} v_{max} \tag{6.51}$$

Then the total lift force for a sphere on the plate electrode is written as

$$L = L_s + L_p \tag{6.52}$$

Then, the force equilibrium equation can be established in the particle migration collection layer as

$$F_E - L - f = 0 \tag{6.53}$$

Where, the drag force is given by

$$f = 3\pi \mu d_p \omega_e \tag{6.54}$$

When equations (6.1), (6.35), (6.52), and (6.54) are substituted into equation (6.53), the migration velocity of the charged particle can be obtained if the shear lift force and the pressure difference lift force are considered as

$$\omega_e = \omega - \frac{L_s + L_p}{3\pi \mu d_p} \tag{6.55}$$

Example 6.4 Using the data of the example 6.3 to calculate the migration velocity of a 5μm particle with relative permittivity of $\varepsilon = 6$. The average gas velocity in a wire-plate ESP is 1m/s, the gas viscosity $\mu = 1.85 \times 10^{-5}$ Pa · s, and the gas density is $\rho = 1.2 \text{kg/m}^3$.

Solution

The data of the example 6.3 are $U = 40$kV, $b = 0.1$m, $r_0 = 1 \times 10^{-3}$ m, $\varepsilon_0 = 8.85 \times 10^{-12}$ C/(V · m), $k = 2.1 \times 10^{-4}$ m²/(V · s), $E_q = 400$kV/m, and $E_p = 206$kV/m

Because of $d_p \geqslant 1$μm, only the field charging is considered. The charge on the particle is

$$q_f = 3\pi \varepsilon_0 E_q d_p^2 \left(\frac{\varepsilon}{\varepsilon + 2} \right) = 6.2 \times 10^{-16} \text{ C}$$

When $d_p \geqslant 1$μm, the Cunningham slip correction factor $C_c = 1$. The theoretical migration velocity of a 5μm particle is

$$\omega = E_p q / 3\pi \mu d_p = 0.11 \text{m/s}$$

When the dust re-entrainment is considered, the Reynolds number is

$$\mathrm{Re} = \frac{\rho d_p v}{\mu} = 0.27$$

From Fig. 2. 5, the lift coefficient C_s is about

$$C_s = 6$$

Then, the shear lift force is

$$L_s = C_s \frac{1}{8}\pi \rho d_p^2 v^2 = 7.06 \times 10^{-11}\ \mathrm{N}$$

The pressure difference lift force is

$$L_p = \frac{1}{21}\pi \rho d_p^2 \left(\frac{d_p}{2b}\right)^{2/7} v_{max}^2 = \frac{1}{21}\pi \rho d_p^2 \left(\frac{d_p}{2b}\right)^{2/7} \left(\frac{8}{7}v\right)^2 = 0.23 \times 10^{-12}\mathrm{N}$$

The real migration velocity of a 5μm particle is

$$\omega_e = \omega - \frac{L_s + L_p}{3\pi \mu d_p} \approx 0.03\mathrm{m/s}$$

It is noticed that the real migration velocity is much smaller than the theoretical migration velocity. Since the effect of the dust re-entrainment is proportional to the square of the gas velocity, the dust re-entrainment can be greatly decreased if the gas velocity is reduced.

6. 5. 4 Specific Collection Area

The specific collection area (S_{CA}) which is defined as the ratio of the total collection area to the total gas flow rate is an important parameter that influences the performance of a ESP. The S_{CA} can be changed by changing either the collection plate area or the gas volume flow rate or both. In effect, changes in S_{CA} result in changes in the treatment time experienced by the particles. Thus, increasing the S_{CA} of a ESP increases the collection efficiency. This relation can be described by Deutsch equation as

$$\eta = 1 - \exp\left(-\frac{A}{Q}\omega\right) = 1 - \exp(-S_{CA}\omega) \tag{6.56}$$

Fig. 6. 14 gives an example of the overall mass collection efficiency of a full-scale ESP collecting fly ash from a coal fired boiler[15]. The ESP had three electrical sections in the gas flow direction, 36 gas passages, and a plate height of 8. 9m. The S_{CA} was varied by changing the boiler load. The temperature and resistivity of the ash ranged from 180℃ to 200℃ and 0. 4Ω · cm to 1. 0×10^{12} Ω · cm, respectively. The curve in Fig. 6. 14 can be used to give a rough evaluation on the performance an ESP in operation. For instance, if the S_{CA} is about 50$\mathrm{m^2/(m^3 \cdot s)}$, the overall collection efficiency should be greater than 99%. Otherwise, the collection performance of the ESP is not quite satisfactory. From Fig. 6. 14, if we want to increase the collection efficiency from 99. 0% to 99. 5%, the S_{CA} have to be increased from 50$\mathrm{m^2/(m^3 \cdot s)}$ to 100$\mathrm{m^2/(m^3 \cdot s)}$.

6. 5. 5 Voltage-current Characteristics

The voltage-current characteristics or *V-I* characteristics is one of an important factors to indicate the performance of an ESP. The higher power consumption means that the higher electric force acts

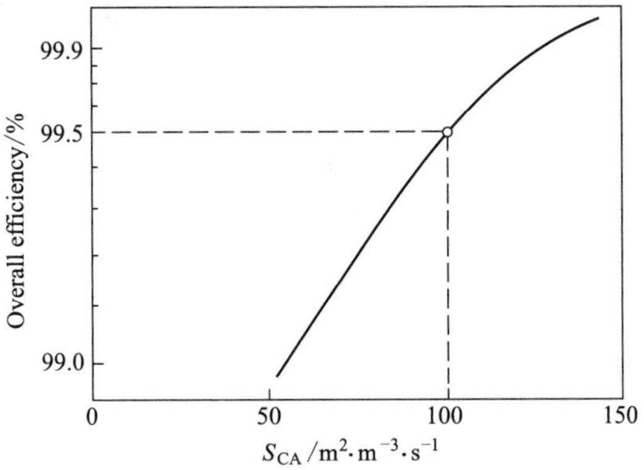

Fig. 6.14 Overall collection efficiency as a function of specific collection area (S_{CA})

on the particles, because the value of the voltage $V \times I$ is the power consumption of an ESP. Therefore, the relative higher collection efficiency can be obtained if the *V-I* characteristics is better.

6.5.5.1 Corona Wire

There are many types of the corona electrodes, such as round wire, star wire or 'quare twisted' wire, barbed wire, etc. Some corona or discharge electrode geometries are shown in Fig. 6.15.

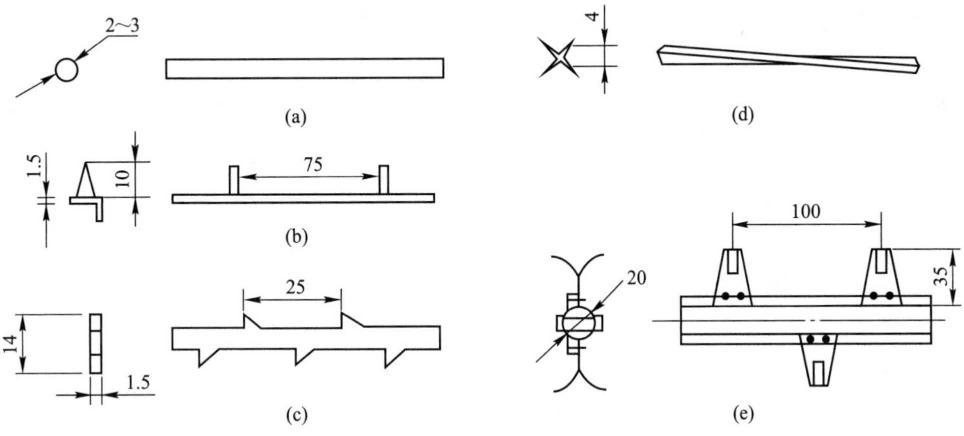

Fig. 6.15 Typical forms of corona electrodes

(a) Round wire; (b) Angle steel barbed wire; (c) Saw tooth wire; (d) Star wire; (e) RS barbed wire

Fig. 6.16 shows the *V-I* characteristics of different types of the corona electrodes schematically. The barbed wire has better *V-I* characteristics because the sharp barb can generate stronger corona discharge under the same condition.

6.5.5.2 Plate Electrode

The plate electrode shape has little effect on the *V-I* characteristics of an ESP. The collection electrodes are the individual grounded surfaces on which particulate matter is collected. Most

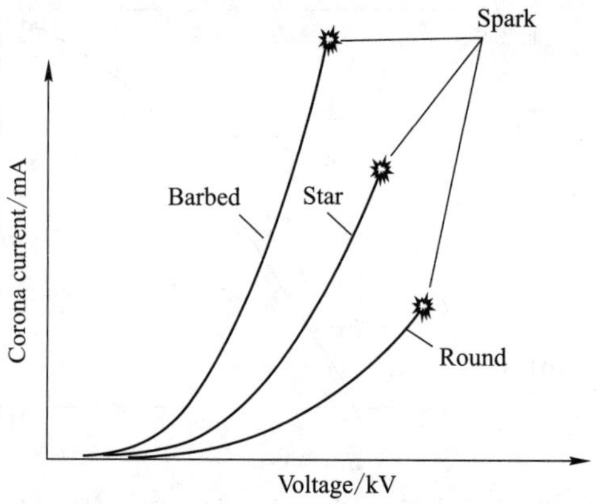

Fig. 6. 16　Schematic *V-I* characteristics of different corona electrodes

collection electrodes are simple configurations, the main considerations being stiffness of the collection plates and shielding of the collected dust layer to prevent dust re-entrainment. An additional requirement is that the edge of the collection plates be free of sharp edge can provide localized high field regions, in case of resulting in sparking at low voltage.

Offset plates are made by bending a flat sheet into a corrugated pattern. The dust precipitated in the troughs is shielded from the main gas stream to minimize the dust re-entrainment. The collection electrodes of ' Z' and ' C' plates, as shown in Fig 6. 17, are commonly used in the industrial application. The plates are usually from 350 ~ 485mm in width and from 3 ~ 15m in height.

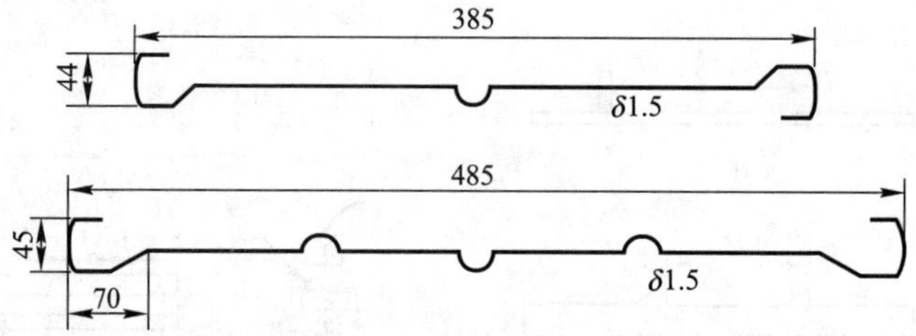

Fig. 6. 17　Configurations of ' Z' and ' C' collection plates

6. 5. 5. 3　Electrode Spacing

The wire to wire distance $2c$ and the plate to plate distance $2b$ do not only affect the spatial distribution of current density, electric field, and space charge density, but also effect the *V-I* characteristics of an ESP greatly.

If the wire to wire spacing is too small, the corona current will become small due to the electric restraining effect. If the wire to wire spacing is too large, the area current density will be reduced. Thus, an optimizing wire distance exists in wire-plate ESP.

White[10] had found experimentally that the corona current is the highest when there were five wires under the geometric condition as shown in Fig. 6. 18.

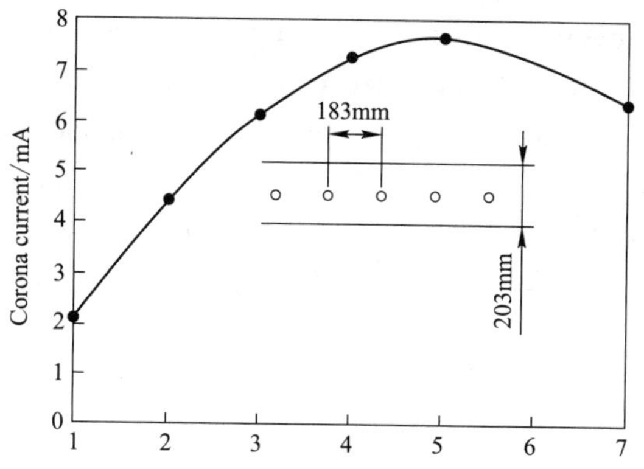

Fig. 6.18 Relation of wire and plate distance

Then, the relation of wire distance and plate distance can be determined by

$$2c/2b \approx 0.9 \tag{6.57}$$

In the conventional ESP, the distance from the plate to plate is about 250mm. The results of the experiments and the practical applications have shown that the collection performance of an ESP can be improved if the plate to plate spacing is increased properly.

The distance from the plate to plate is often ranged from 350~500mm. It is call the wide plate spacing ESP. The advantages of the wide plate spacing ESP include that (1) the ESP cost is reduced because less steel material is used, (2) the back corona phenomenon caused by the dust resistivity will be weaken to a certain extent, and (3) the collection efficiency may be increased because the average electric field at the plate is higher for the same average current density at the plate.

6.5.6 Dust Resistivity

The appropriate dust resistivity for the ESPs ranges from $1\times10^4\ \Omega\cdot$cm to $5\times10^{10}\ \Omega\cdot$cm. However, in many cases, the useful operating current density in an ESP is limited by the resistivity of the collected particulate layer[16]. If the resistivity of the collected particulate layer is sufficiently high, electrical breakdown of the layer will occur at the value of the current density which in most cases is undesirably low. Depending on the value of the applied voltage, either a condition of sparking or the formation of stable back corona that will be detrimental to the ESP performance should be avoided.

The mechanism of the back corona of high dust resistivity can be described by Ohm's law as

$$E_R = \rho_R J \geqslant E_a \tag{6.58}$$

Where E_R——the electric field strength in dust layer in V/m;

ρ_R——the dust resistivity in $\Omega\cdot$m;

J——the current density in A/m^2;

E_a——the breakdown field strength ($E_a = 3\times10^6$ V/m in standard conditions).

Equation (6.58) means that if the electric field strength in dust layer E_R is greater than the electric breakdown field strength E_a, the back corona in dust layer will occur. It is clear that if the dust resistivity is high, the current density has to be reduced which leads to the ESP collection efficiency to be reduced in order to avoid back corona. The curve in Fig. 6. 19 demonstrates how sensitive performance is to the resistivity of the collected fly ash. For this particular situation, an increase in resistivity from $10^{10}\,\Omega \cdot cm$ to $5\times10^{11}\,\Omega \cdot cm$ will result in a decrease in overall mass collection efficiency from 98. 1% to 81%. This example points out why a knowledge of the resistivity of the collected ash layer is crucial in designing an ESP.

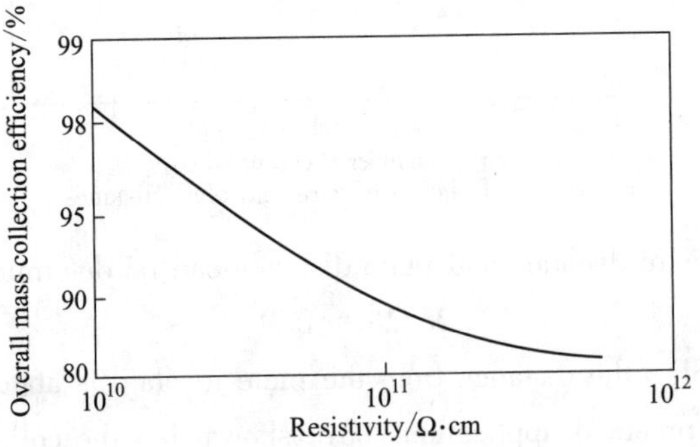

Fig. 6. 19　Effect of resistivity on overall mass collection efficiency

There are many factors influencing the dust resistivity: the gas temperature, the humidity, and the chemical composition. The dust resistivity of a collected layer of fly ash varies with temperature is illustrated in Fig. 6. 20. Above about 200℃, resistivity decreases with increasing temperature and is independent of flue gas composition. Below about 100℃, resistivity decreases with decreasing temperature and is dependent upon moisture and other constituents of the flue gases.

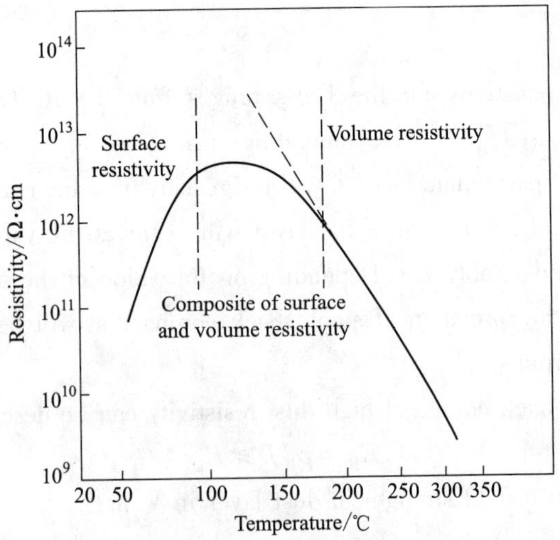

Fig. 6. 20　Temperature-resistivity relationship for fly ash

When temperature is higher than 180℃, the conductivity of the particulate layer (volume conduction) is increased[17]. However, in order to reduce the dust layer resistivity it is obviously not reasonable to increase temperature above 200℃. Then, we can make use of the surface conduction property of the collected dust layer. Dust resistivity can be reduced by gas conditioning immediately before precipitation. This can be achieved by injecting small quantities of gas conditioning agents. Each of the chemicals used as a conditioning agent has varying results depending on the process. Chemicals used include a hot ammonia-air mixture, triethylamine-air mixture, steam, and water vapor. The conditioned gas must be insured to be above the dew point temperature.

6.5.7 Gas Distribution

The uniform velocity distribution in an ESP is one of the assumptions in deriving the Deutsch equation. However, in fact, the gas flow is not uniform because the cross sectional area is changing from the inlet duct to the upper inlet ESP. This will lead to the nonuniform gas velocity distribution in the electric field of the ESP[18]. If some flow devices, such as screens, are not installed in the front of the first electric field, the nonuniform gas velocity will worsen the ESP collection performance due to uneven treatment of particles in different velocity zones.

The criteria of evaluating the velocity distribution uniformity is given by

$$\sigma_v = \sqrt{\frac{1}{n}\sum_{i=1}^{n}\left(\frac{v_i - v}{v}\right)^2} \qquad (6.59)$$

Where v_i ——the velocity at the measuring point;

n——the number of the measuring points;

v ——average velocity of an ESP.

If $\sigma_v \leqslant 0.1$, the gas velocity distribution is excellent. If $0.1 < \sigma_v \leqslant 0.15$, the gas velocity distribution is fine. If $1.5 < \sigma_v \leqslant 0.25$, the gas velocity distribution is acceptable.

The distance between the measuring points is less than 1m. The measuring sections are selected in the front of the first electric field and in the rear of the last electric field. A cross sectional area is divided into smaller equal area. Each measuring point is located in the center of the small area. If the height of the collection electrode plate is less than 8 m, the number of the measuring points should be at least 8 in the vertical direction of the measuring section.

In order to get anacceptable velocity distribution, the perforated plates are used. Usually, three perforated plates are installed in the front of the first electric field. While in the rear of the last electric field, only one or two perforated plates are needed.

6.6 ESP Design

6.6.1 Information Required for an ESP Design

So far the general principles of the ESP have been discussed. Attention has been directed to the dry ESP which depends on the dust being dry, and covers such applications as power stations, cement works, and metallurgical refineries. Even though the applications of the ESPs are versatile, the basic information required for an ESP design which should be presented as follows, is almost no different for all ESPs in industrial applications.

6. 6. 1. 1 Volume Flow Rate and Gas velocity

The volume flow rate of the flue gas is one of the important performance indexes of an ESP. When the flow rate Q is known, the section area A_s can be determined as

$$A_s = Q/v \tag{6.60}$$

Where the average velocity v in the electric field is usually selected as about 1m/s in an industrial ESP design.

6. 6. 1. 2 Gas Temperature and Composition

According to the temperature range, the ESPs are classified into cold-side ESP (<200℃) and hot-side ESP (300~450℃). It is more economical to cool the hot gases preferably to near 300℃. The 450℃ limit is the strength limit of the carbon steel. There are two main advantages of the hot-side ESP operation. One is that the fly ash resistivity is getting much lower due to the volume conductivity at high temperature. The other aduanage is that the heat exchanger is not needed for cooling the high temperature gases if the temperature in the range between 300℃ and 450℃. However, some technical measures have to be taken to prevent the expansion of the materials inside the ESP. In order to ensure the ESP operation reliability and the lifetime, the gas temperature is often reduced below 200℃. Of course, the gas temperature must be kept above the dew point temperature.

Only when the gas temperature is known, the gas viscosity and density can be determined since these two physical quantities are necessary in ESP design. The relationships of the gas dynamic viscosity and density with the gas temperature are respectively given by

$$\mu = \mu_0 \frac{273 + C}{T + C}\left(\frac{T}{273}\right)^{1.5} \tag{6.61}$$

$$\rho = \rho_0 \frac{T_0 p}{T p_0} \approx 1.2 \times \frac{273}{273 + T} \tag{6.62}$$

Where μ_0——the gas dynamic viscosity at the standard conditions, $\mu_0 = 1.85 \times 10^{-5}$ Pa · s;

C——a constant, $C = 122$;

T——the gas temperature in K.

Attention should be paid on the gas composition, such as SO_x, NO_x, CO_2, O_2, N_2, and H_2O. If the contents of SO_x (SO_2 and SO_3), H_2O, and O_2 are relatively higher, an ESP is preferable for high resistivity dust collection and corona generation. If the content of N_2 is high, the corona may be affected.

6. 6. 1. 3 Inlet Dust Concentration and Particle Size Distribution

The inlet dust concentration must be known for an ESP design. The inlet dust concentration up to 10g/Nm^3 has been successfully treated without problems in electric field. If the inlet dust concentration exceeds 30g/Nm^3, the corona could possibly be blocked in the first electric section of the ESP. It is better to install a pre-cleaning separator, such as a multiple cyclone, in front of the ESP if the dust concentration in the gas is too high.

The particle size distribution which had been discussed in Chapter 1 is very important in an ESP design[19]. Particularly the mass median diameter $d_{p_{50}}$ of the inlet particulate must be obtain.

The dust composition also has a certain effect on the collection performance. Especially the dust resistivity has a strong effect on the collection performance of the ESP. It is easy to measure the dust resistivity at room condition, but not at high temperature. If the information on the dust resistivity in the gas is not available, a designer should investigate the similar processes to get the data from existing ESPs or roughly estimate the dust resistivity according to the trend illustrated in Fig. 6.20.

6.6.2 Overall Mass Collection Efficiency

In an ESP design, the particulate collection requirement must be given first. The particulate pollutant control requirement usually is the emission mass concentration, or the emission standard. Thus, the overall mass collection efficiency of the ESP is required by

$$\eta_R = 1 - \frac{c_e}{c_i} \tag{6.63}$$

Where η_R ——the overall mass collection efficiency of the ESP which meets the needs of the emission mass concentration;

c_e ——the emission mass concentration;

c_i ——the inlet mass concentration of the flue gas.

Therefore, the ESP design must be satisfied with

$$\eta_D \geqslant \eta_R \tag{6.64}$$

Where η_D ——the overall mass collection efficiency of the designed ESP.

However, the Deutsch equation (6.36) or (6.37) is the fractional efficiency equation. The relation between the overall collection efficiency and the fractional efficiency is given by

$$\eta_D = \sum \Delta F_i \eta_i \tag{6.65}$$

Where ΔF_i —— the particle mass fraction.

If the particle mass frequency distribution f is known, the relation between the overall mass collection efficiency and the fractional efficiency can be expressed by

$$\eta_D = \int_0^{\infty} f \eta \, \mathrm{d} d_p \tag{6.66}$$

However, the numerical calculation has to be used to find the overall collection efficiency from equation (6.66). This calculation procedure is awkward and tedious.

For an approximate prediction of the overall collection efficiency, an empirical overall collection efficiency had been given by Zhao and Pfeffer[20] as

$$\eta_D = \eta(d_{p_{50}}) = 1 - \exp\left[- \frac{A}{Q} \omega(d_{p_{50}}) \right] \tag{6.67}$$

Where $d_{p_{50}}$ ——the particle mass median diameter.

Because the $d_{p_{50}}$ is often larger than 1 μm, from equations (6.6) and (6.35), the migration velocity of the particle with the median diameter in equation (6.67) is given by

$$\omega(d_{p_{50}}) = \frac{\varepsilon_0 \varepsilon}{\mu(\varepsilon + 2)} E_p E_q d_{p_{50}} \tag{6.68}$$

Thus, it is very convenient to use equation (6.67) to calculate the overall collection efficiency for an ESP design because the mass median diameter $d_{p_{50}}$ of the initial particulate is easy to be obtained.

Because the idealized conditions corresponding to the Deutsch equation seldom exist in actual practice, industrial ESP design is often based on the empirical migration velocities, called the effective migration velocity, for use in the Deutsch equation which is given by

$$\eta_D = 1 - \exp\left(-\frac{A}{Q}\omega_e\right) \tag{6.69}$$

Where ω_e ——the effective migration velocity of the particulate, which is determined in Table 6.1.

Table6.1 Effective migration velocities of different particulates

Particulate	Effective migration velocity/m · s^{-1}	Particulate	Effective migration velocity/m · s^{-1}
Fly ash	0.08~0.12	Limestone powder	0.04~0.05
Cement	0.09~0.10	Magnesium	0.04~0.05
Sintering dust	0.06~0.20	Zinc oxide	0.04
Iron oxide	0.07~0.23	Gypsum	0.19~0.20
Tar particulate	0.03~0.05	Aluminum oxide	0.08

The effective migration velocities in Table 6.1 are the 'average' data from existing installations of ESPs. The values are varying with the ESP performance improvement due to the technology development. Therefore, even though the equation (6.69) is a simplest formula to calculate the overall collection efficiency for an ESP design, it is suggested that it would be better to use equation (6.69) just for checking purpose.

6.6.3 Design Procedure

After the initial information: the flow rate, gas temperature, particle size distribution, inlet particle concentration, and required emission concentration, has been obtained, an ESP can be designed. In order to describe the design procedure clearly, an example is given as follows.

Example 6.5 Design an ESP for fly ash emission control in a coal combustion boiler. The original parameters are given as follows.

Gas volume flow rate $Q = 500\text{km}^3/\text{h} = 140\text{m}^3/\text{s}$, emission standard $c_e = 20\text{mg/m}^3$, inlet concentration $c_i = 2000\text{mg/m}^3$, gas temperature $T = 400\text{K}$, gas permittivity $\varepsilon_0 = 8.85 \times 10^{-12}\ \text{C/(V} \cdot \text{m)}$, relative permittivity $\varepsilon = 6$, $d_{p_{50}} = 10\mu\text{m}$, $b = 0.2\text{m}$, $c = 0.15\text{m}$, voltage supply $U = 80\text{kV}$, $k = 2.1 \times 10^{-4}\text{m}^2/(\text{V} \cdot \text{s})$.

Solution

(1) Migration velocity. The average charging electric field strength is

$$E_q = \frac{U}{b} = 400\text{kV/m}$$

The field strength at the edge of the corona regain E_0 is

$$E_0 = 3 \times 10^6 f(\delta + 0.03\sqrt{\delta / r_a}) = 2.6 \times 10^6 \text{V/m}$$

where, $f = 0.7$, $\delta = T_0 p / T p_0 = 0.68$, $r_a = 0.002$m.

The voltage U_0 of triggering the corona is

$$U_0 = r_a E_0 \ln(b / r_a) = 24\text{kV}$$

The current density is

$$i = \frac{8\pi\varepsilon_0 k}{b^2 \ln(b / r_a)} U(U - U_0) = 0.113 \times 10^{-5} \text{A/m}$$

The deviation σ is

$$\sigma = \frac{\pi r_a E_0}{\sqrt{2\pi\left(\frac{r_a^2 E_0^2}{b^2} + \frac{i}{2\pi\varepsilon_0 k}\right)}} = 0.43$$

Form equation (6.47), the collection electric field strength is

$$E_p = E(b, 0) = \frac{\pi r_a E_0}{\sqrt{2\pi}\sigma}\left[1 + 2\exp\left(-\frac{c^2}{\sigma^2}\right)\right] = 36.6\text{kV/m}$$

The gas viscosity at $T = 400$K is

$$\mu = \mu_0 \frac{273 + C}{T + C}\left(\frac{T}{273}\right)^{1.5} = 2.48 \times 10^{-5} \text{Pa} \cdot \text{s}$$

The migration velocity of the particulate with the median diameter 10μm is

$$\omega(d_{p_{50}}) = \frac{\varepsilon_0 \varepsilon}{\mu(\varepsilon + 2)} E_p E_q d_{p_{50}} = 0.097\text{m/s}$$

Comparing this value with the effective migration velocity of fly ash in Table 6.1, the migration velocity of 0.097m/s is acceptable.

(2) Total collection area. The required overall mass collection efficiency of the ESP is

$$\eta_R = 1 - \frac{c_e}{c_i} = 99\%$$

According to equation (6.66), the collection area is

$$A = -\frac{Q \ln(1 - \eta_R)}{\omega(d_{p_{50}})} = 6.6\text{km}^2$$

(3) Collection plate height. If the gas velocity in electric field is taken as 1m/s. the total section area is

$$F = \frac{Q}{v} = 140\text{m}^2$$

If the total section area $F' \geqslant 80\text{m}^2$, two inlet should be designed. Then each inlet section area is

$$F' = 70\text{m}^2$$

The collection plate height is

$$h = \sqrt{F'} = 8.4\text{m}$$

If $h<8$m, the increment is 0.5m. If $h>8$m, the increment is 1.0m. Thus, the collection plate height is

$$h = 9\text{m}$$

(4) Number of the gas passages in ESP. The number of passages in each inlet are

$$N' = \frac{F'}{2bh} = 19$$

The total number of passages are

$$N = 38$$

(5) Total length of electric field is

$$L = A/2Nh = 9.6\text{m}$$

(6) Number of electric fields. In order to achieve better rapping acceleration, the length of the collection electrode usually ranges from 3m to 4m. Therefore, several shorter electric fields are arranged in series in an industrial ESP. In this example, the number of the electric fields is

$$n = L/l = 3$$

The length of each field is

$$l = 3.2\text{ m}$$

Because the ESPs have been commercially produced, an ESP can be chosen directly according to above design.

Exercises

6.1 A fly ash particle receives 1×10^{-16}C saturation charge in a 4.0 kV/cm charging electric field. Calculate the particle diameter [$\varepsilon_0=8.85\times10^{-12}$ /(CV · m), $\varepsilon=6$].

6.2 A 5μm spherical particle is moving at a velocity of 3cm/s in a 3.5 kV/cm uniform electric field. Calculate the charge on the particle (the dynamic viscosity μ at standard condition is 1.85×10^{-5}Pa · s).

6.3 A horizontal parallel wire-plate ESP consists of a single passage 4m high×6m long, with a 0.3m plate-to-plate spacing. A collection efficiency of 95% is obtained with a flow rate of 4200m^3/h, and the inlet concentration is 5g/m^3. Calculate the following:

(1) the average gas velocity in the electric field;

(2) the outlet concentration;

(3) the effective drift (migration) velocity for this system;

(4) a revised collection efficiency if the flow rate increased to 6000m^3/h;

(5) a revised collection efficiency if the plate-to-plate spacing decreased to 0.2m.

6.4 A horizontal parallel wire-plate ESP consists of 10 passages 5m (high) ×10m (long), with a 0.4m plate-to-plate spacing to treat 72000m^3/h of gas containing 5g/m^3 of carbon black particulate. Using the average migration velocity for the system to calculate the collection efficiency of the unit. Assume the carbon black particle size distribution may be represented by a median particle diameter of 10μm.

6.5 Calculate the field strengths at y = 0m, 0.075m, and 0.15m respectively for the unit described in example 6.2.

6.6 The collection plate area of an old ESP was $A_1 = 2000m^2$. The collection efficiency of this ESP η_1 was 95%, which was satisfied the old emission standard $C_1 = 100mg/m^3$. But now the new Emission Standard is $C_2 = 50mg/m^3$. How much area of a new ESP is needed to meet the need of new emission standard?

6.7 It is known that an ESP performance can be upgraded by changing the electric power supply to raise the operating voltage. Suppose the original efficiency was 95% and the original voltage was 40kV. If the voltage is increased to 50kV and everything else is kept constant, what is the estimated new collection efficiency?

6.8 A gas stream contains particles of sizes which are 10μm, 7μm, and 3μm. The particle density is the same for all three sizes, and the mass concentration in the gas stream of each of the three sizes of particles is the same ($G_0 = G = G_3 = 0.333$). We pass this gas stream through an ESP that obeys equation (6.37). If the overall collection efficiency of the ESP is 95%, what are the individual collection efficiency for each of the three sizes of particle?

6.9 An ESP has three identical chambers operating in parallel. When the flow is distributed equally (one third to each), the particle collection efficiency is 95%. Now, as a result of misdistribution, the flows become 50%, 30%, and 20% to the three chambers. The total flow is unchanged. What is the overall particle collection efficiency under this flow condition?

References

[1] Mizuno A. IEEE transactions on dielectrics and electrical insulation [J]. Electrostatic precipitation, 2000, 7 (5): 615-624.

[2] Jaworek A, Krupa A, Czech T, et al. Modern electrostatic devices and methods for exhaust gas cleaning: A brief review [J]. Journal of Electrostatics, 2007, 65 (3): 133-155.

[3] Jaworek A, Marchewicz A, Sobczyk A T, et al. Two-stage electrostatic precipitators for the reduction of $PM_{2.5}$ particle emission [J]. Progress in Energy and Combustion Science, 2018, 67: 206-233.

[4] Talaie M R. Mathematical modeling of wire-duct single-stage electrostatic precipitators [J]. Journal of Hazardous Materials, 2005, 124 (1): 44-52.

[5] Dramane B, Zouzou N, Moreau E, et al. Electrostatic precipitation in wire-to-cylinder configuration: Effect of the high-voltage power supply waveform [J]. Journal of Electrostatics, 2009, 67 (2): 117-122.

[6] Chang J, Lawless P A, Yamamoto T, et al. Corona discharge processes [J]. IEEE Transactions on Plasma Science, 1991, 19 (6): 1152-1166.

[7] Mcdonald J R, Smith W B, Spencer H W, et al. A mathematical model for calculating electrical conditions in wire-duct electrostatic precipitation devices [J]. Journal of Applied Physics, 1977, 48 (6): 2231-2243.

[8] Lami E, Mattachini F, Sala R, et al. A mathematical model of electrostatic field in wires-plate electrostatic precipitators [J]. Journal of Electrostatics, 1997, 39 (1): 1-21.

[9] Oglesby S Jr, Nichols G B. *Electrostatic Precipitation* [M]. New York: Marcel Dekker, 1978.

[10] White H J. *Industrial Electrostatic Precipitation* [M]. New York: ddison-Wesley, 1963.

[11] Cooperman P. A new theory of precipitator efficiency [J]. Atmospheric Environment, 1971, 5 (7): 541-551.

[12] Lin G, Chen T, Tsai C, et al. A modified deutsch-anderson equation for predicting the nbanoparticle collection efficiency of electrostatic precipitators [J]. Aerosol and Air Quality Research, 2012, 12 (5):

697-706.

[13] Robinson M. A modified deutsch efficiency equation for electrostatic precipitation [J]. Atmospheric Environment, 1967, 1 (3): 193-204.

[14] Zheng C, Liang C, Liu S, et al. Balance and stability between particle collection and re-entrainment in a wide temperature-range electrostatic precipitator [J]. Powder Technology, 2018, 340: 543-552.

[15] McDonald J K. *Electrostatic Precipitator Manual* [M]. New York: Noyes, 1982.

[16] Wiggers H. *Measurement of dust resistivity-Back corona in electrostatic precipitators* [C]. IEEE Powertech Conference, 2007, 87 (3): 93-96.

[17] Barranco R, Gong M, Thompson A, et al. The impact of fly ash resistivity and carbon content on electrostatic precipitator performance [J]. Fuel, 2007, 86 (16): 2521-2527.

[18] Enliang L, Yingmin W, Raper J A, et al. Study of gas velocity distribution in electrostatic precipitators [J]. Aerosol Science and Technology, 1990, 12 (4): 947-952.

[19] Clack H L. Particle size distribution effects on gas-particle mass transfer within electrostatic precipitators [J]. Environmental Science & Technology, 2006, 40 (12): 3929-3933.

[20] Zhao Z, Pfeffer R. A simplified model to predict the total efficiency of gravity settlers and cyclones [J]. Powder Technology, 1997, 90 (3): 273-280.

7 Fabric Filtration

Fabric filtration can be defined as the process of separating dispersed particles from a gas by means of porous fabric media. The fabric filters are frequently used to remove solid particles from industrial gases, whereby the dusty gas flows through fabric bags and the particles accumulate on the cloth.

The fabric filters, especially the membrane filters, can separate the particulate with very high efficiency. Therefore, the fabric filters have become a major class of particulate air pollution control devices today[1].

The performances of a fabric filter are collection efficiency, pressure drop, air-cloth ratio, and bag life[2]. Thus, we begin with an analysis of the mechanisms of collection by a single cylindrical fiber placed in a particulate-laden gas flow. Then we consider the overall collection efficiency of a fibrous filter. Next, we will give theoretical expressions on the pressure drop of the filter bed or bag. Finally, we will discuss the fabric materials, industrial fabric filters, and design for applications.

7.1 Mechanisms of Particle Capture by a Fiber

Based on the separation mechanism, the fabric filtration media can be divided into two basic types of filters, surface and depth[3]. Woven fabrics are intended to be surface filters. Membrane filters are typical surface filters. Felts appear as depth filters.

New or over-cleaned fabrics presents usually depth filtration function at beginning, and penetration of some extremely fine particles occurs until a dust cake is formed. Thus, for the clean woven fabrics, the particles are collected by fibers rather than by dust cake on the surface of the woven fabrics. When the dust cake is formed on the surface of the filter, the filtration function of the dust cake is thought to be surface filtration[4].

In order to obtain the overall collection efficiency of a fibrous filter, we must establish the theoretical filtration models of a single fiber. A fiber can be taken as an isolated cylinder normal to the gas flow. Three distinct mechanisms can be identified whereby particles in the gas reach the surface of the cylinder: diffusion, interception, and inertial impaction, as shown in Fig. 7.1.

Collection may also result from electrostatic attraction when either particles or fiber or both possess a static charge[5]. These electrostatic forces may be either direct, when both particle and fiber are charged, or induced, and only one of them is charged. Such charges are usually not present unless deliberately introduced during the manufacture of the fiber. We will discuss the mechanisms of the electrostatic enhancement effects in Chapter 9.

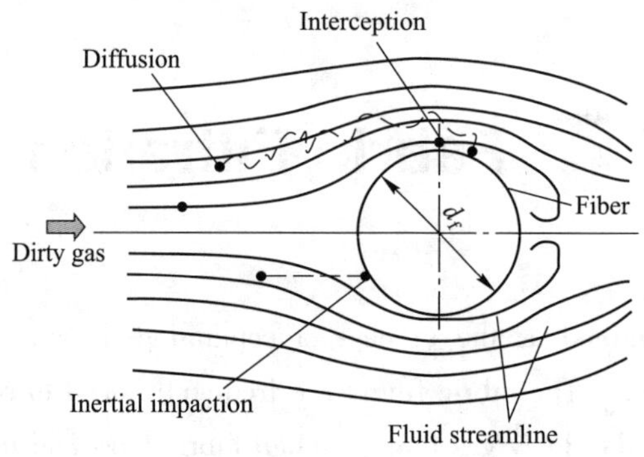

Fig. 7. 1 Classical mechanisms of particle collection by a fiber

7. 1. 1 Interception

Interception takes place when a particle, flowing the streamlines of flow around a cylinder. Thus, if the streamline on which the particle center lies is within a distance $d_p/2$ of the cylinder, interception occurs. This streamline is the path of the particle. Collection by interception can be approximated by neglecting any particle inertia and assuming that incoming particles simply follow the streamlines of the flow exactly, as shown in Fig. 7. 2.

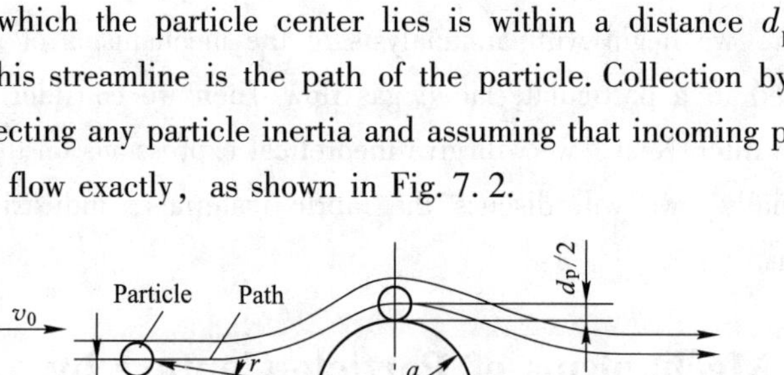

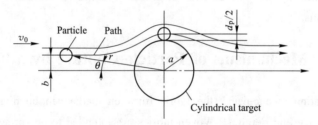

Fig. 7. 2 Deposition of particles on a cylinder by interception

It is clear that the interception mechanism depends on the flow field around a cylinder placed normal to the flow. The Reynolds number, based on the cylinder diameter d_f, is written as

$$\mathrm{Re} = \frac{\rho d_f v_0}{\mu} \tag{7.1}$$

If $\mathrm{Re} \gg 1$, the flow field belongs to the potential flow. The stream function is given by

$$\psi = v_0 \left[1 - \left(\frac{a}{r}\right)^2\right] r\sin\theta \tag{7.2}$$

Where v_0——the gas velocity far from the cylinder

a——the radius of the cylinder, $d_f = 2a$.

When $\theta = \pi/2$, $r = a + d_p/2$, from equation (7. 2), we have

$$v_0\left[1 - \left(\frac{a}{r}\right)^2\right] r\sin\theta = v_0\left[1 - \left(\frac{a}{a + d_p/2}\right)^2\right](a + d_p/2) \tag{7.3}$$

When, $r \to \infty$, in Fig. 7. 2, $r\sin\theta = b$, equation (7. 3) becomes

$$b = a + d_p/2 - \frac{a^2}{a + d_p/2} \tag{7.4}$$

When both sides of equation (7.4) are divided by a, the interception efficiency in the potential flow is obtained by

$$\eta_R = \frac{b}{a} = 1 + \frac{d_p}{2a} - \frac{1}{1 + d_p/2a} = 1 + G - \frac{1}{1 + G} \tag{7.5}$$

Where G——the interception parameter, which is given by

$$G = d_p/d_f \tag{7.6}$$

If the Reynolds number $Re \leqslant 1$, the viscous flow can be assumed. The stream function is expressed as

$$\psi = \frac{v_0}{2L_a}\left[2\ln\left(\frac{r}{a}\right) - 1 + \left(\frac{a}{r}\right)^2\right] r\sin\theta \tag{7.7}$$

Where L_a——Lamb constant, which is given by

$$L_a = 2 - \ln Re \tag{7.8}$$

The interception efficiency in the viscous flow can be developed from equation (7.7) as

$$\eta_R = \frac{1}{L_a}\left[(1 + G)\ln(1 + G) - \frac{G(2 + G)}{2(1 + G)}\right] \tag{7.9}$$

For the flow in the fabric filtration, the Reynolds number is usually of order unity or smaller. Therefore, the viscous flow model is applicable in the fabric filtration.

7.1.2 Inertial Impaction

Inertial impaction occurs when particle is unable to follow the rapidly curving streamlines around the a cylinder, and continues to move toward the cylinder because of its inertia. Thus, collision occurs because of the momentum of the particle. Inertial impaction is important for particles exceeding 1μm[6].

Unlike in the interception in which the particle remains in its streamline, the particle trajectory in the inertial impaction is difficult to be determined by its motion equation. The mechanism analysis indicates that the inertial impaction efficiency is the function of the Stokes number. The Stokes number is given by

$$St = \frac{\rho_p d_p^2 v_0}{18\mu d_f} \tag{7.10}$$

For potential flow, a single fiber collection efficiency attributable to inertial impaction is expressed empirically as

$$\eta_I = \frac{St^3}{St^3 + 0.77St^2 + 0.22} \tag{7.11}$$

Equation (7.11) adequately fits most of the experimental data.

In viscous flow, the inertial impaction effect is very small in the fabric filtration because the slow gas velocity leads to the Reynolds number $Re < 1$.

7.1.3　Diffusion

Diffusion is a capture mechanism resulting from the Brownian motion of a particle caused by multiple collisions with the gas molecules. only with small particles (<0.5μm) is this mechanism significant. As the particle size decreases, the diffusion effect increases. When the particle size is smaller than 0.1μm, the diffusion collection efficiency may exceed 50%, while other capture mechanisms tend to zero, as shown in Fig. 7.3.

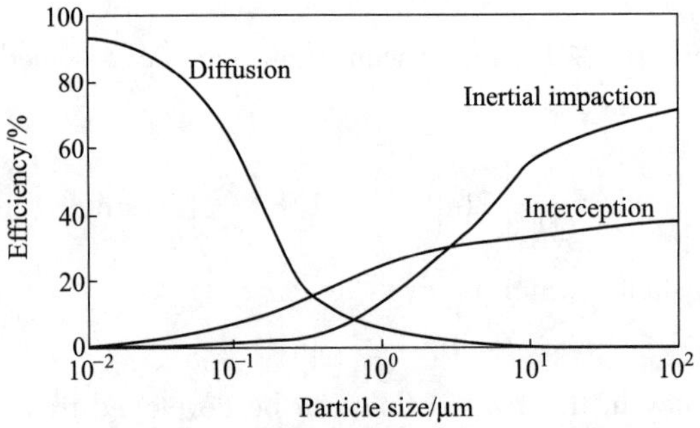

Fig. 7.3　Efficiencies of different mechanisms of a single fiber

The Peclet number, Pe, relates fluid motion to the diffusion coefficient and hence is a useful parameter in calculating diffusion collection efficiency written as

$$Pe = \frac{d_p v_0}{D} \tag{7.12}$$

where, the diffusion coefficient D is determined by equation (3.2).

In the fabric filtration, the Reynolds number and the Peclet number are $Re<1$, and $Pe > 1$. In this case, an empirical equation is given by[6]

$$\eta_D = C \frac{1}{(La)^{1/3}} Pe^{-2/3} \tag{7.13}$$

Where　C——the experimental coefficient.

For $d_p < 0.1\mu m$, $C = 1.75$ (Langmuir), and for $0.1 \leqslant d_p < 1.0$, $C = 2.92$ (Natason).

7.1.4　Combination Collection Efficiency of Independent Mechanisms

Even though the collection efficiencies of three independent mechanisms listed above are frequently such that one mechanism is dominant in a particular range of particle size, for the approximate calculation of the combination effects, an acceptable overall collection efficiency of combined these three independent mechanisms can be expressed as

$$\eta = 1 - (1 - \eta_R)(1 - \eta_I)(1 - \eta_D) \tag{7.14}$$

In order to control the fine particle emission, the filtration velocity is very small in the fabric filtration application. Therefore, the inertial impaction mechanism is negligible. Thus, the interception and diffusion mechanisms for a single fiber are considered as

$$\eta = 1 - (1 - \eta_R)(1 - \eta_D) \tag{7.15}$$

7.2 Collection Efficiency of a Fibrous Filter Bed

7.2.1 Depth Filtration Efficiency of a Filter

A fibrous filter bed is viewed as a loosely packed assemblage of single cylinders. Even though the fibers are oriented in all directions in the bed, from a theoretical point of view, the bed is treated as if every fiber is normal to the gas flow through the bed.

For a clean fabrics, suppose the solid fraction of the filter is β, the fiber diameter is $d_f = 2a$, the cross sectional area of the filter is A, the thickness is L, the upstream gas velocity of the filter is v_0, and the concentration of the particle-laden gas is c_0.

We consider the balance over a slice of thickness dh, as shown in Fig. 7.4.

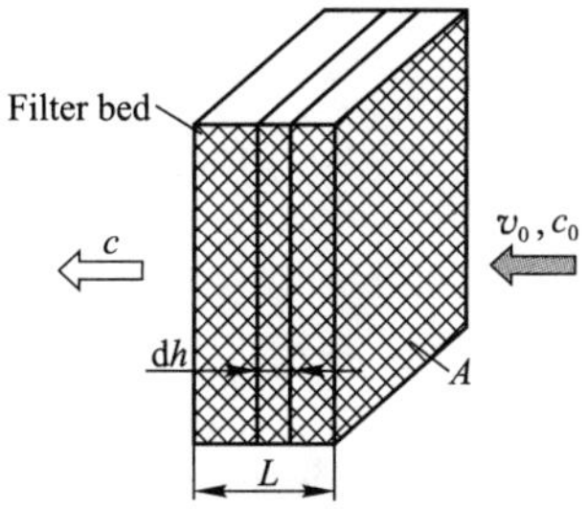

Fig. 7.4 Particle collection in clean filter

The total length L_f in the volume of A dh is given by

$$L_f = \beta A \mathrm{d}h / \pi a^2 \tag{7.16}$$

In the element dh, the particle concentration is c. Because the collection efficiency of a single fiber is η, then the mass of the collected particles in the element dh is given by

$$\mathrm{d}m = 2avL_f c\eta = 2ac\eta v\beta A \mathrm{d}h / \pi a^2 \tag{7.17}$$

Where v——the gas velocity in the filter, which given by

$$v = v_0 / (1 - \beta) \tag{7.18}$$

Then equation (7.17) becomes

$$\mathrm{d}m = 2c\eta \frac{v_0 \beta A \mathrm{d}h}{(1 - \beta)\pi a} = \frac{4c\eta v_0 \beta A \mathrm{d}h}{\pi(1 - \beta) d_f} \tag{7.19}$$

When the particle-laden gas flows through the element dh, the reduced mass of the particles is given by

$$\mathrm{d}m = -Av_0 \mathrm{d}c \tag{7.20}$$

Let equation (7.19) is equal to equation (7.20), we have

$$\frac{\mathrm{d}c}{c} = -\frac{4\beta\eta \mathrm{d}h}{(1 - \beta)\pi d_f} \tag{7.21}$$

When integrated over a filter of length L, subject to the concentration from c_0 to c, equation (7.21) is given by

$$\frac{c}{c_0} = \exp\left[-\frac{4\beta\eta L}{\pi(1-\beta)d_f}\right] \tag{7.22}$$

Thus, the collection efficiency of a clean filter is expressed as

$$\eta_0 = 1 - \frac{c}{c_0} = 1 - \exp\left[-\frac{4\beta\eta L}{\pi(1-\beta)d_f}\right] \tag{7.23}$$

Example 7.1 Compute the collection efficiency of the felt filter for 1μm particle. The calculating data are given as $L = 3$mm, $d_f = 30$μm, $\beta = 0.3$, $v_0 = 1.5$m/min, $\mu = 1.85 \times 10^{-5}$ Pa · s, and $\rho = 1.2$kg/m^3.

Solution

The inertial impaction can be neglected because the particle diameter is only 1μm.

(1) Interception. The Reynolds number flow around the cylindrical fiber is

$$\mathrm{Re} = \frac{\rho d_f v_0}{\mu} = 0.05$$

Lamb constant is

$$L_a = 2 - \ln\mathrm{Re} = 5$$

The interception parameter is

$$G = d_p/d_f = \frac{1}{30}$$

From equation (7.9), the interception efficiency is

$$\eta_R = \frac{1}{L_a}\left[(1+G)\ln(1+G) - \frac{G(2+G)}{2(1+G)}\right] = 8.9 \times 10^{-4}$$

(2) Diffusion. The diffusion coefficient is for spherical particles suspended in a perfect gas, D is estimated from the kinetic of gases as

$$D = k_B T C_c / 3\pi\mu d_p = 2.7 \times 10^{-11}$$

where, $k_B = 1.38 \times 10^{-23}$ J/K.

The Peclet number is

$$\mathrm{Pe} = \frac{d_p v_0}{D} = 9.3 \times 10^2$$

The diffusion collection efficiency is calculated by equation (7.13) as

$$\eta_D = C\frac{1}{(\mathrm{La})^{1/3}}\mathrm{Pe}^{-2/3} = 0.018$$

where, $C = 2.92$ (Natason).

(3) Combination efficiency. From equation (7.15), The combination efficiency of interce-ption and diffusion is

$$\eta = 1 - (1-\eta_R)(1-\eta_D) = 0.018$$

(4) Collection efficiency of the filter for 1μm particle. From equation (7.23), the collection efficiency of the filter for 1μm partide is

$$\eta_0 = 1 - \exp\left[-\frac{4\beta\eta L}{\pi(1-\beta)d_f}\right] = 62.3\%$$

7.2.2 Surface Filtration Efficiency of a Filter

After inside of a fibrous filter bed has contained enough particulate, the particles in the oncoming gas stream will deposits on the surface of the filter. Thus, as the amount of the collected particles increases, the dust cake is formed, as shown in Fig. 7.5. This dust cake becomes the filter which acts as a sieve leading to the collection efficiency increasing[7]. Actually, the time of the depth filtration is very short. The surface filtration plays main role during the filtration process. Especially the membrane filters, which have been widely used in particulate pollutant control today, are typical surface filters[8].

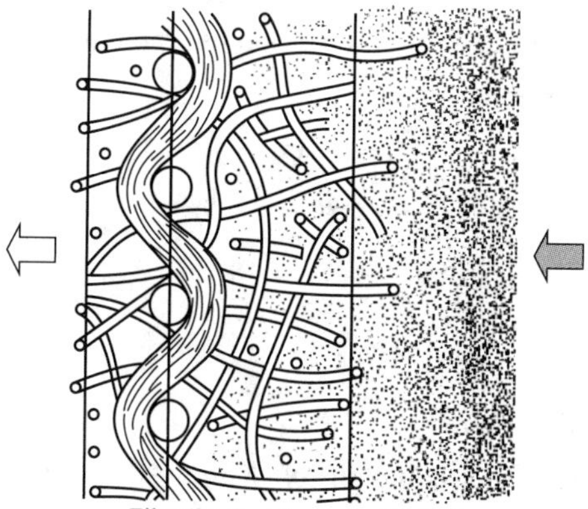

Fig. 7.5 Dust cake formed on the surface of the filter

The basic concepts of the surface filtration are: (1) all the particles could not get through the filter medium if the diameter of the particles larger than the pore of the filter, (2) the collected particles are cumulated on the surface of the filter to form the dust cake, (3) the gas is cleaned by the dust cake, not by the filter, and (4) both the efficiency and pressure drop are increased with the time[9].

The surface filtration means that the dust in the gas is collected by the dust cake on the surface of the filter. The mechanism analysis process is similar to the depth filtration. If we assume that the particles deposited on the surface of the filter bed are spherical and the same size[10], the collection efficiency of a single layer of spheres with porosity can be developed based on the collection efficiency of a spherical target (a particle), which will be discussed in Chapter 8. At last, the overall collection efficiency calculation equation of the dust cake can be obtained. Even though the collection efficiency of the surface filtration can be given theoretically, the theoretical result has less reference in bag-house design and application. By contraries, the pressure drop variation during the filtration of a filter is more important in practice.

7.3 Pressure Drop of a Fibrous Filter Bed

7.3.1 Pressure Drop of a Clean Filter Bed

It is very useful to determine the pressure drop variation during the filtration of a filter because it is related to the power consumption and filter cleaning time[11]. In industrial filter, the gas velocity in the individual pores is so slow that the flow is laminar. Therefore, Darcy formula, the well-known relation for laminar flow of a fluid in a porous medium, can be used to describe the pressure drop of a gas flowing through a filter bed, written as

$$\Delta p_1 = \frac{1}{k_1}\mu v L \tag{7.24}$$

Where k_1——the permeability of a porous medium;

L——the thickness of a porous medium;

v——given by equation (7.18) in the filter bed .

Because k_1 is different from the different porous mediums, for the fabric filtration, a dimensionless parameter X is introduced as

$$X = \frac{a^2 \Delta p_1}{\mu v L} = \frac{a^2}{k_1} \tag{7.25}$$

or

$$\Delta p_1 = \frac{\mu v L}{a^2} X \tag{7.26}$$

Where a——the radius of the fiber.

According to the pressure drop equation of Kozeny-Carman, an empirical dimensionless parameter X was given by Sullivian Hertel[6], which is given by

$$X = \frac{22\beta^2}{(1-\beta)^3} \tag{7.27}$$

Brinkman pointed that equation (7.27) is only suitable for the solid fraction $\beta > 0.12$. Davies also proposed an empirical equation of X from experiments which may be more reasonable in practice. That is

$$X = 16\beta^{3/2}(1 + 56\beta^3) \tag{7.28}$$

If the dimensionless parameter X is determined, the pressure drop of the clean filter can be calculated.

Example 7.2　At the standard condition, the gas viscosity $\mu = 1.85 \times 10^{-5}$ Pa · s, the gas velocity coming to the filter bed is $v_0 = 0.05$m/s, the solid fraction of the filter is $\beta = 0.2$, the thickness of the filter bed is $L = 2$mm, and the diameter of the fiber is $d_f = 2a = 30\mu$m. Calculate the pressure drop of the clean filter bed.

Solution

From Sullivian equation (7.27), the dimensionless parameter X is

$$X = \frac{22\beta^2}{(1-\beta)^3} = 16.3$$

From Davies equation (7.28), the dimensionless parameter X is

$$X = 16\beta^{3/2}(1 + 56\beta^3) = 18.5$$

We can see that the results calculated by Sullivian and Davies equations do not have much difference. From equation (7.26), the pressure drop of the clean filter bed is

$$\Delta p_1 = \frac{\mu v L}{a^2} X = \frac{\mu v_0 L}{a^2(1-\beta)} X = 223 \sim 253\text{Pa}$$

7.3.2 Pressure Drop of a Filter Bed with Cake Filtration

When the dust-laden gas flows through a filter bed, the particles will deposit inside or on the surface of the filter. In this case, the total pressure drop is the combination of the pressure drop of the fabric and the dust cake. That is

$$\Delta p = \Delta p_1 + \Delta p_2 \tag{7.29}$$

Where Δp_1 ——the pressure drop of clean filter bed which is determined by equation (7.26).

The pressure drop of the fabric and the dust cake can also be described by Darcy formula as

$$\Delta p_2 = \frac{1}{k_2} \mu v_2 H \tag{7.30}$$

Where k_2——the permeability of the dust cake;

v_2 ——the velocity inside of the dust cake;

H——the dust cake thickness.

The velocity inside of the dust cake v_2 is expressed as

$$v_2 = v_0 / \varepsilon_p \tag{7.31}$$

Where ε_p ——the dust cake porosity.

Assume the collection efficiency of the filter is nearly 100%. The dust cake thickness H is

$$H = Vc_0 / \rho_p (1 - \varepsilon_p) A \tag{7.32}$$

Where V——the gas volume flowing through the filter during time t;

c_0 ——the particle concentration of the gas;

ρ_p ——the particle density;

A ——the filtration area.

Because of $V = v_0 A t$, equation (7.32) becomes

$$H = \frac{v_0 c_0}{\rho_p (1 - \varepsilon_p)} t \tag{7.33}$$

Substituting equation (7.33) into equation (7.30), the pressure drop of the dust cake filtration is

$$\Delta p_2 = \frac{\mu}{k_2 \rho_p \varepsilon_p (1 - \varepsilon_p)} c_0 v_0^2 t \tag{7.34}$$

In equation (7.34), the permeability of the dust cake k_2 is difficult to estimated accurately because k_2 is related with the particle size distribution, particle shape, etc. Therefore, according to equations (7.26) and (7.34), a modified form of the pressure drop of a filter bed with the dust cake filtration is given by

$$\Delta P = \zeta_1 L v_0 + \zeta_2 c_0 v_0^2 t \tag{7.35}$$

Equation (7.35) is more applicable because the specific resistance coefficient of the fabric ζ_1 and the specific resistance coefficient of the dust ζ_2 are easy to be determined by experiments.

Example 7.3 Establish an experimental formula of a big-house. The experimental data include that approach gas velocity is $v_0 = 1\text{m/min}$, the thickness of the filter bed is $L = 2\text{mm}$, and the fly ash concentration in gas is 3g/m^3. When the filtration begins, the pressure drop of the cleaned bag is 200Pa. After 10 min, the pressure drop of the bag increases to 800Pa.

Solution

At $t = 0$, the specific resistance coefficient of the bag is

$$\zeta_1 = \Delta p_1 / L v_0 = 100$$

At $t = 10\text{min}$, the specific resistance coefficient of the dust cake is

$$\zeta_2 = \frac{\Delta p_2}{c_0 v_0^2 t} = \frac{\Delta p - \Delta p_1}{c_0 v_0^2 t} = 20$$

Then, the pressure drop of a big-house in the filtration process is

$$\Delta P = 100 L v_0 + 20 c_0 v_0^2 t$$

Where, the gas velocity is in m/min, the thickness of the filter is in mm, the concentration is in g/m^3, and the filtration time is in min.

7.4 Filter Media

7.4.1 General Description

The natural fabrics are seldom used in industrial air pollution control. The synthetic fabrics are more widely used in particulate separation[12].

The membrane filters, such as the polyester felt or glass fabric felt covered by a thin layer of the Teflon membrane, can be used to effectively collect the fine particles in all dry gases if the gas temperature is less than 180℃[13,14]. In the most cases, the emission concentration is less than 10mg/m^3 or even less than 5mg/m^3 when the membrane fabric is used for bags in a industrial bag-house. Therefore, the application of the membrane filters is becoming more and more popular in the particulate pollutant control in order to satisfy the emission standards.

When the gas temperature ranges from 180℃ to 350℃, the fibreglass fabric should be used. If the gas temperature exceeds 350℃, for example, the integrated coal gasification combined cycle (IGCC) power generation system [operation temperature 350 ~ 1000℃ and pressure 10 ~ 25atm ($1\text{atm} = 1.01 \times 10^5\text{Pa}$) in particulate control system], the porous ceramic filter has to be used[16].

Even though the particle-laden gases at high temperature can be treated directly, the control system cost of the maintenance expense are quite high. Usually some measures are taken to cool down the high temperature gases below 350℃. In the most case, the temperature of the gas at the end of a production system is often less than 180℃, then a lot of common synthetic fabrics can be used. In order to meet the needs of the emission standard, the membrane filters and cartridge filters are widely used today.

7.4.2 Membrane Filter

There are three main kinds of membranes, etched or leached nuclepore polycarbonate membrane, sintered membrane, and stretched membrane made from PTFE (polytetrafluorethylene).

The thickness of the membrane is often less than 0.1mm. The pore size can be accurately prepared by manufacture process. The membrane filters can be used for microfiltration and ultrafiltration, or even reverse osmosis depending on the pore size of the membrane. But for industrial air pollution control, the microfiltration membrane filters with the pore size range from 0.1μm to 5μm are only needed to separate the particulate in gases. The membrane pore size selection depends on the particle size distribution.

A polycarbonate membrane, as shown in Fig. 7.6, is formed by means of radioactive impinging into narrow tracks and solution etching into pores. The porosities of polycarbonate membrane filters are usually quite small. Thus, polycarbonate membrane filters are seldom used in industrial air pollution control. They can be used in viruses and bacteria separation since the concentration is usually low.

The sintered membrane, as shown in Fig. 7.7, has relatively uniform pore size and high porosity. The manufacture cost of the sintered membrane is higher than that of the stretched membrane.

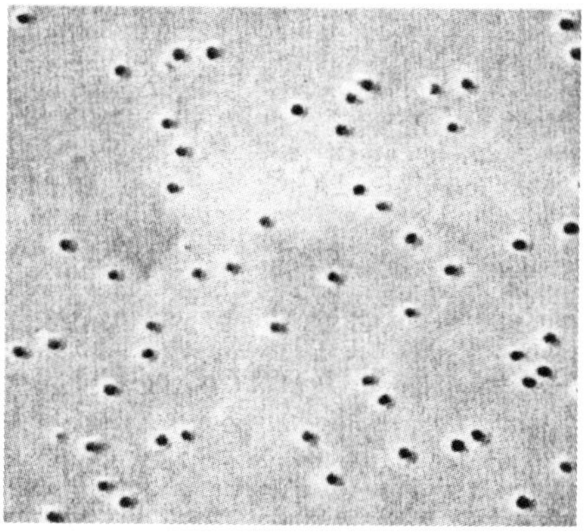

Fig. 7.6 Nuclepore polycarbonate membrane

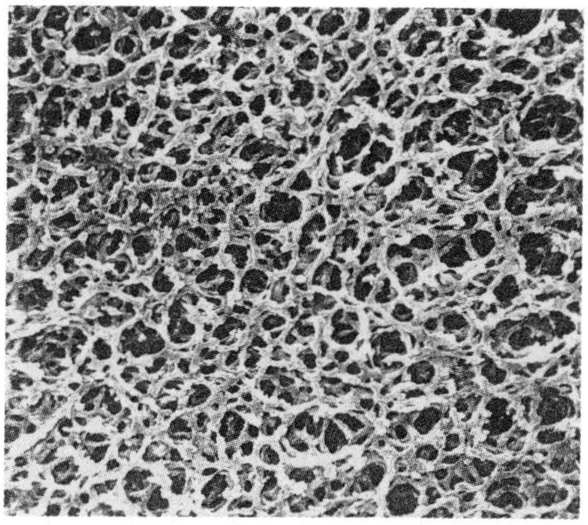

Fig. 7.7 Sintered membrane (PTFE)

The stretched Teflon membrane filters, as shown in Fig. 7.8, are more applicable in aerosol particulate separation because of the cheaper price, longer lifetime, and wider pore size range[17]. The Teflon (polytetrafluorethylene, PTFE) membrane filters that find such wide use in air filtration are manufactured by controlled stretching of dense Teflon films. This process has been patented by W. L. Gore and Associates, and the filters are marketed by them in large sheets under the name Gore-Tex[18]. Gore-Tex membrane filters are supplied in pore size of 0.02μm, 0.2μm, 0.45μm, 1.0μm, 3.0μm, 5.0μm, and 10~15μm.

The porous Teflon membrane can be laminated to any of the man-made fabrics, creating a substantially different surface for that fabric and influencing many of the fabric filtration properties. One can convert. For example, as shown in Fig. 7.8, the polyester felt fabric to the Teflon membrane fabric filter that the Teflon membrane is coated on the surface of the polyester felt.

The Teflon fabric has the properties of high temperature rating, excellent corrosion resistance,

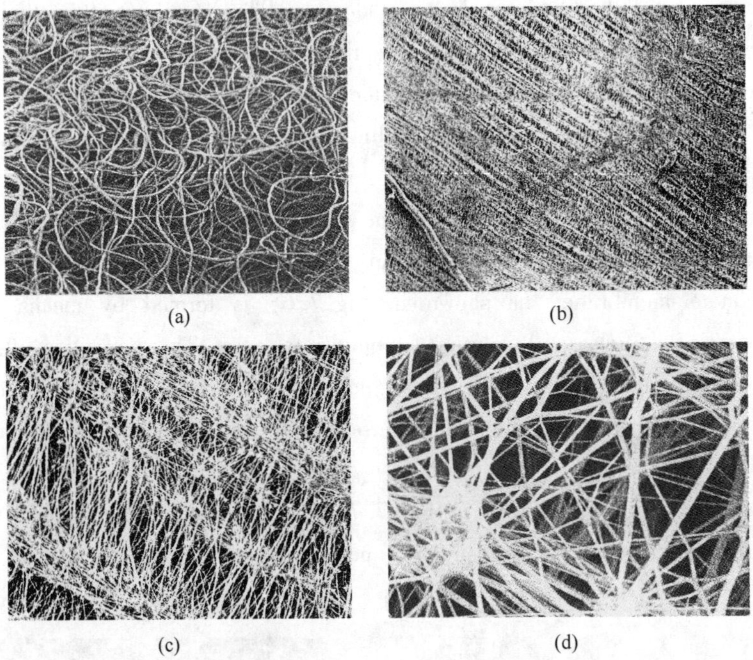

Fig. 7. 8　Gore-Tex membrane on polyester felt

(a) Polyester felt (15×); (b) 60×; (c) 600×; (d) 6000×

and good flex life. Even though the Teflon membrane promotes surface deposition, affects cake release and alters the drag characteristics, it is clear that the properties of the supporting fabrics are important in gas filtration. The main properties of the fabrics are temperature, acid and alkali resistance. The properties of some common fabrics for dry filtration are given in Tables 7. 1 and 7. 2 respectively.

Table 7. 1　Properties of fabrics for low and medium temperature

Fiber \ Generic name / Trade name	Cotton —	Polymid Nylon 66	Polypropylene Herculon	Polyester Dacron	Acrylic copolymer Orlon	Homopolymer acrylic Draylon T
Recommended continuous operation temperature /℃	82	94	94	132	120	140
Maximum operation temperature /℃	94	121	107	150	120	140
Specific density	1. 50	1. 14	0. 90	1. 38	1. 16	1. 17
Resistance alkalies	Good	Good	Excellent	Fair	Fair	Fair
Resistance to mineral acids	Poor	Poor	Excellent	Fair	Good	Very good
Resistance to organic acids	Poor	Poor	Excellent	Fair	Good	Excellent
Resistance to oxidizing agents	Fair	Fair	Good	Good	Good	Good
Resistance to organic solvents	Good	Very good	Excellent	Good	Very good	Very good

Table 7.2 Properties of fabrics for high temperature

Fiber / Generic name / Trade name	Aramid	Glass	PTFE	Polyphenylene sullide	Polybenzimidazole	Ceramic
	Nomex	Fiberglass	Teflon	Flylon	PBI	Nexlet 312
Recommended continuous operation temperature /℃	204	260	260	190	260	1150
Maximum operation temperature /℃	232	290	290	232	290	1427
Specific density	1.39	2.54	2.30	1.38	1.43	2.7
Resistance alkalies	Good	Fair	Excellent	Excellent	Good	Good
Resistance to mineral acids	Fair	Very good	Excellent	Excellent	Excellent	Very good
Resistance to organic acids	Fair	Very good	Excellent	Excellent	Excellent	Very good
Resistance to oxidizing agents	Poor	Excellent	Excellent	Poor	Fair	Excellent
Resistance to organic solvents	Good	Very good	Excellent	Excellent	Excellent	Excellent

Tables 7.1 and 7.2 are the useful guide reference to choose the fabrics reasonably. Actually, we could not simply say 'good' or 'bad' fabrics. If the temperature is low, it is not necessary to choose the high temperature resistant fabrics since the high temperature resistant fabrics are often expensive. Otherwise, the gas properties must be understood clearly in order to choose the fabrics correctly.

In the conventional filtration, the textile fabrics listed in Tables 7.1 and 7.2 are used directly as the filtration medium. After the Teflon membrane was invented, the membrane filtration technology is developed when fabrics (felt) are treated by the membrane and used for industrial dirty gas cleaning. The collection performances of bag-house have been improved greatly[19]. The comparison of the conventional filtration and the membrane filtration is shown schematically in Fig. 7.9 when polyester felt and Teflon membrane (median pore size 1.0μm) filter are used in particulate filtration. It is indicated that all particles larger than pore size 1.0μm will be retained on the surface of the membrane filter by a sieving mechanism.

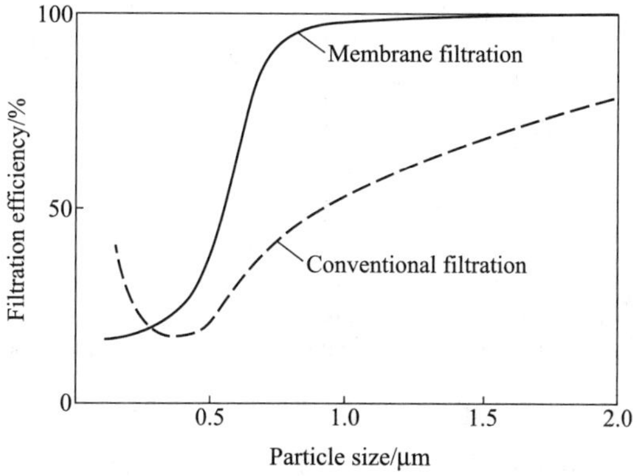

Fig. 7.9 Collection efficiency comparison of conventional filtration and membrane filtration

The advantages of membrane filtration include that: (1) the filter can be manufactured in a uniform and highly precise manner, (2) because of its high porosity, the filter supports high flow rates of the fluid (higher gas-cloth ratio), and (3) membrane filters can function as sieves, permitting the separation of particles of different sizes.

Membrane filtration totally changed the fibrous filtration manner. It converts the depth filtration (the combination of interception, inertial impaction, diffusion) into the surface filtration (sieving). Thus, the filtration mechanisms are simplified by membrane filtration. The purpose of the particulate separation is that an air pollution control system is hoped to be operated stably and reliably under the condition of satisfying the emission standard. In the most industrial particulate pollutant control applications, the membrane filtration can achieve this purpose.

7.5 Industrial Bag-house

The basic differences in fabric filters are methods of cleaning the collected dust from the filter cloth and the configuration of cloth within the collector housing. Generally the cleaning methods are commonly employed, such as pulse jet, mechanical shaking, backwash (on reverse flow), and sonic. These are listed in order of their popularity and are briefly described below.

7.5.1 Pulse Jet

Pulse jet is commonly referred as a continuous cleaning filter, as shown in Fig. 7.10. The system utilizes felted fabrics in the form of an envelope or circular bag, supported on an internal frame. Thus, dust filtration takes place on the outside of the element with the clean air passing out through the centre.

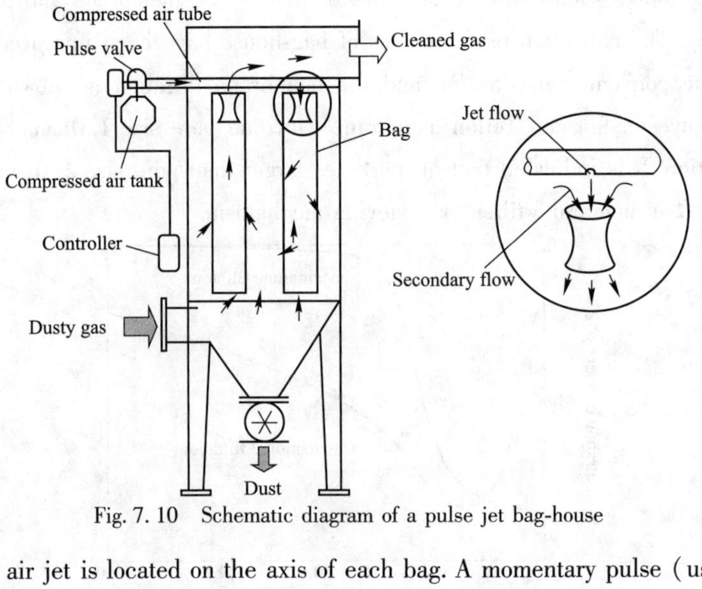

Fig. 7.10 Schematic diagram of a pulse jet bag-house

A compressed air jet is located on the axis of each bag. A momentary pulse (usually from 0.1s to 0.2s) of high pressure (from 0.4MPa to 0.7MPa) compressed air inflates the bag to distends the cloth like a balloon.

The significant advantages of this system are better bag cleaning, lower bag erosion, and higher gas to cloth ratio[20].

7.5.2 Mechanical Shaking

The mechanical shaking bag-house is one of the oldest bag-houses still in application, as shown in Fig. 7.11. The generally circular bags are suspended from the top, and fastened to a 'thimble' plate at the bottom.

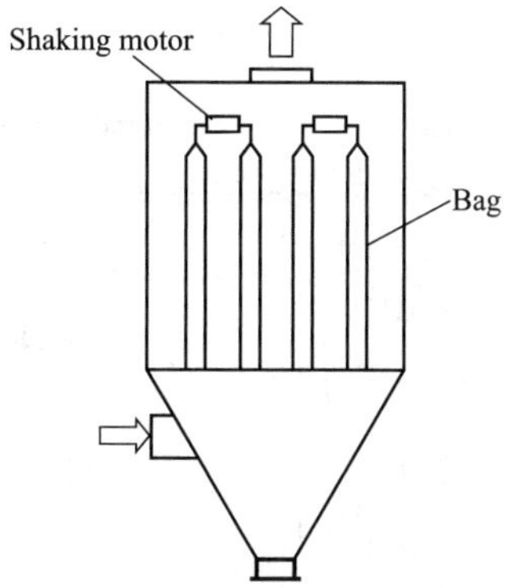

Fig. 7.11 Schematic diagram of a mechanical shaking bag-house

The gas flow is from the inside to outside of the bags, such that on shaking or vibrating at the top, the collected dust falls into the hopper below[21].

The system is simple but has the disadvantages of very large size due to the low air to cloth ratios, and short bag life time due to the strong shaking force.

7.5.3 Backwash

The backwash is also commonly known as back pressure, or collapse-reverse flow cleaning. In every case, filter flow is shut off, and an air flow (reverse cleaning pressure flow) is established in the opposite direction, i.e., from the clean side to the dirty side of the bags. Therefore, there are two types of backwash bag-house, inside filtration and outside filtration, as shown in Fig. 7.12.

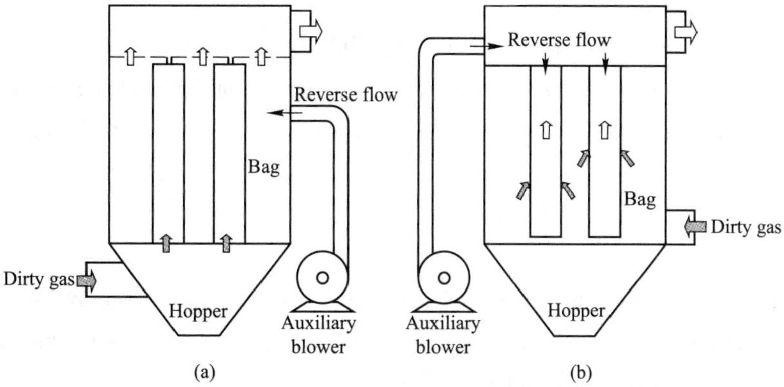

Fig. 7.12 Schematic diagram of the backwash bag-house

(a) Inside filtration; (b) Outside filtration

Bag tension and backwash pressure are critical. Insufficient suspension tension makes collapse easier. The maximum reverse pressure across the bags is about 50mm water gauge (500Pa). The spacer rings are sewn into the bag at 800mm to 1200mm intervals from top to bottom to make backwash cleaning more effective.

Simple, gentle reverse air flow through the cloth is an ineffective cleaning mechanism. Therefore, backwash cleaning is often used with other cleaning means, shaking or sonic methods.

In order to improve the cleaning effect, a circle backwash bag-house has been developed, as shown in Fig. 7. 13.

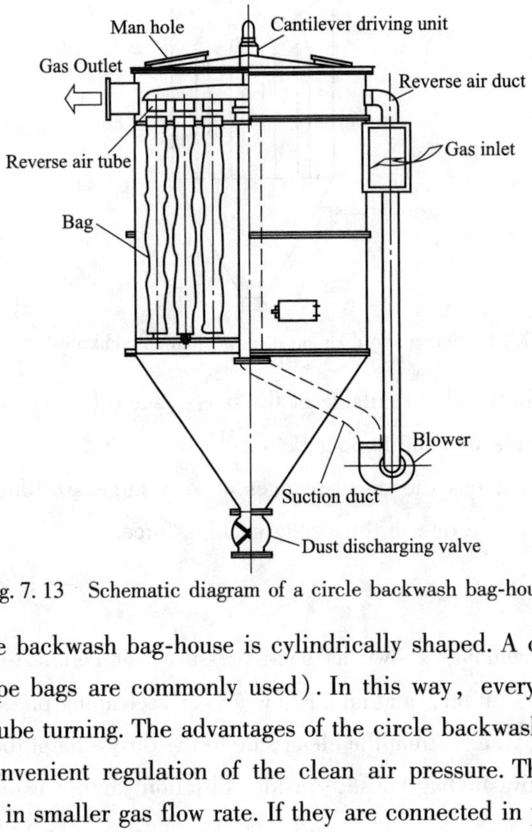

Fig. 7. 13 Schematic diagram of a circle backwash bag-house

The body of the circle backwash bag-house is cylindrically shaped. A cantilever tube is turning above the bags (envelope bags are commonly used). In this way, every bag is cleaned in each circle of the cantilever tube turning. The advantages of the circle backwash are low consumption of the cleaning air and convenient regulation of the clean air pressure. The circle backwash bag-houses are usually used in smaller gas flow rate. If they are connected in parallel, the flow rate of millions cubic meter per hour can be treated.

For the smaller gas flow rate (less than $100km^3/h$), the blower can be put on the top of the bag-house. The blower sucks the cleaned gas directly from the outlet chamber of the bag-houses. Thus, the pressure loss is reduced because the suction and reverse ducts become shorter, as shown in Fig. 7. 14.

7. 5. 4 Sonic

The sonic bag cleaning system is applied to a conventional collectors, usually as a substitute for other cleaning methods, such as shaking. In other words, filter flow must be valved off the compartment being cleaned as with other methods.

Sound is heard outside of the casing and minor vibration can be felt. Bags do not appear to

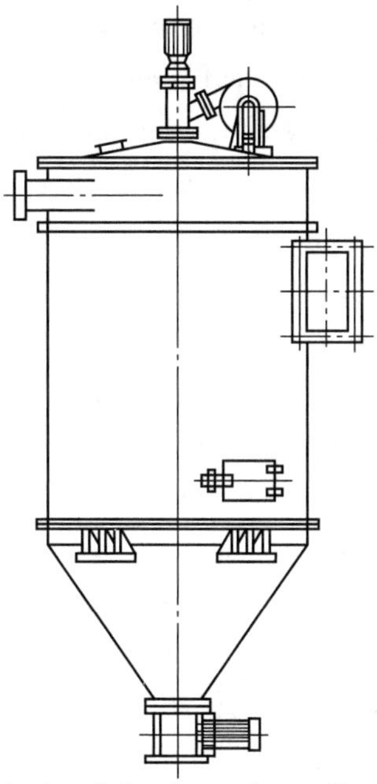

Fig. 7.14 Circle backwash bag-house for small gas flow rate control

move. Therefore, the damage to bags is very low. Sonic cleaning is a good adding method to improve the cleaning effect[22], such as connection with backwash cleaning.

7.6 Bag-house Design

7.6.1 Filter Emissions

One of the most important data for bag-house design is the filter emission. Most pulse jet filters will have emissions below the currently general emission standard of 20mg/Nm^3 (Chapter 1). By careful selection of filtration velocities, filtration media, and application of the pulse cleaning system, it is possible to achieve emissions below 10mg/Nm^3 within five years. It is anticipated that emissions could be reduced to less than 5mg/Nm^3 if the Teflon membrane fabric is applied.

7.6.1.1 Relation of Emission and Bag Resistance

Consideration of the relationship between emissions and filter bag resistance (pressure drop) is also required. Most dusts fortunately do not exhibit emissions proportional to filter bag resistance. But unfortunately there are some dusts which do. The latter are usually free flowing, non agglomerating types, alumina being a classic example. To avoid high emission levels with these dusts, it is necessary to reduce filtration velocities or select filtration media which resists particulate movement though the fabric.

7.6.1.2 Relation of Emission and Pulse Cleaning

The pulse cleaning will also effects on bag emission. The emission concentration during pulse may be more than three times higher than the emission in filter operation.

Even though the cleaning time is very short, which does not increase the average emission greatly, over bag cleaned or too long cleaning time interval should be avoided. Most pulse jet filters require 8 to 10 liters (free air) per m^2 per pulse to achieve filter fabric acceleration forces in the range 200~300g. The determination of pulse cleaning frequency depends on the dust cake resistance.

7.6.1.3 Relation of Emission and Inlet Gas

The particle concentration, size distribution, and gas distribution do effect the bag emission. The gas curried with coarse particles may lead to filter bag abrasion. When membrane filter is used, the filter bag abrasion must be avoided. For all bag-house designs, the gas velocity in the filter bag compartment must be reduced to no more than 2m/s. The gas flow must be distributed uniformly over the filter bags[23]. If the inlet particle concentration is quite high ($\geq 15g/Nm^3$), lower filtration velocity should be selected.

7.6.2 Filtration Velocity

The value of the filtration velocity and the air to cloth ratio is same. The unit of filtration velocity is usually in m/min, where the air to cloth ratio is in $m^3/(min \cdot m^2)$.

The selection of filtration velocity is probably the most difficult aspect of fabric filter design. The filtration velocity determination is usually based on prior experiences, similar installations, and selection guides. The selected filtration velocity values are conservative because the nature of the industrial dust-laden gas is very complex. For example, the suggested filtration velocity for shaking is 0.5~0.9m/min, for backwash (reverse flow cleaning) is 0.4~0.8m/min, and for pulse jet is 0.8~1.2m/min.

If the filtration velocity or the air to cloth ratio is determined, the area of the fabric can be calculated according to the gas flow rate. Then the number of the bags can be calculated according to the area of the filtration media, bag length and diameter. The ranges of the bag length, diameter, and the center distance between two bags are given in Table 7.3.

Table 7.3 Bag length, diameter, and center distance between two bags for filter design

Cleaning	Bag diameter ϕ/mm	Bag length l/m	Bag distance e /mm
Mechanical shaking	100~200	<3	1.6ϕ
Backwash (inside filtration)	200~300	2~12	1.4ϕ
Pulse jet (outside filtration)	120~200	2~8	1.5ϕ

The space occupation of the bag-house can be calculated empirically by

$$S = 1.5 \frac{A}{\pi \phi l} e^2 \tag{7.36}$$

Where S——the land area in m^2;

A——the total area of fabric in m^2;

e——the center distance between two bags in m;

l——the bag length in m;

ϕ——the bag diameter in m.

Example 7.4 Select a bag-house for collecting the particulate from a discharging point of the belt conveyor. The basic data and requirements are given in Table 7.4.

Table 7.4 Basic data of for particulate control of a discharging point of the belt conveyor

Gas temperature/℃	RH/%	Flow rate/$m^3 \cdot h^{-1}$	Particle median size/μm	Concentration/$g \cdot m^{-3}$	Emission/$mg \cdot m^{-3}$
30~60	70~90	80000~108000	5	5	30

Solution

(1) Bag cleaning method. Since the fluctuation of the gas flow rate is great and particle size is quite small, a pulse jet cleaning is selected in order to satisfy the emission requirement.

(2) Bag-house size. For pulse jet bag-house, the filtration velocity is about 1.0 m/min. Then the total filtration area is

$$A = Q/v = 1330 \sim 1800\text{m}^2$$

If the size of the bag is selected as ϕ160mm×6000mm, from Table 7.3, the distance between bags is $e = 1.5\phi$. From equation (7.36), the land space occupation of the bag-house is

$$S = 1.5\frac{A}{\pi\phi l}e^2 = 37 \sim 51\text{m}^2$$

(3) Bag-house type. The CD series pulse jet bag-house is selected from the bag-house guide. The technical parameters are given in Table 7.5.

Table 7.5 Technical parameters of CD104-9/18 pulse jet bag-house

Filtration area /m^2	Filtration velocity /$m \cdot min^{-1}$	Flow rate /$km^3 \cdot h^{-1}$	Bag size /mm×mm	Inlet concentration /$g \cdot m^{-3}$	Bag number	Land space /mm×mm	Jet pressure /MPa	Bag-house resistance /Pa
1944	≤1.03	≤120	ϕ160×6000	≤20	648	9440×4040	0.2~0.3	≤1500

7.6.3 Cleaning Design

To achieve the optimum cleaning effects for filter bags is very difficult. The problem of bag cleaning is still one of the most significant researches.

7.6.3.1 Backwash

A distinguish advantage of the backwash is that the fiberglass can be used for bag at higher gas temperature because the backwash is a gentle cleaning action. The typical configuration of a fabric filter bag used in backwash bag-house is shown in Fig. 7.15.

A tensioning mechanism is required for tensioning the bags to a predetermined force level in

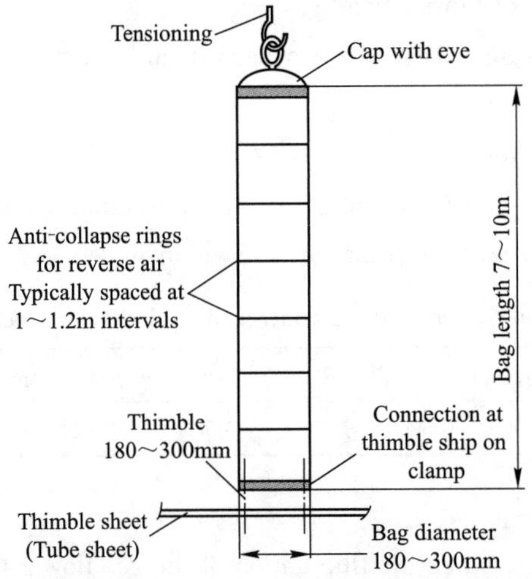

Fig. 7. 15　Typical configuration of a fabric filter bag for reverse-air flow cleaning

order to clean efficiently and with minimum bag wear. With suitable tension (0. 2～0. 5kg/cm), the bag collapses into a multi-lobed star-shaped cross section under the reverse-air flow. This pattern both cracks the accumulated dust cake and allows it to fall to the hopper. Too little tension allows the bag to collapse into a flat pancake shape which cracks the cake less effectively and partially blocks both the flow of reverse air and the passage of released cake. These low tension actions reduce cleaning efficiency and bag life sufficiently that common practice is to sew anti-collapse rings into bag for the express purpose of minimizing those deleterious consequences of operating with non-ideal tension. Anti-collapse rings add significant cost to the bag, but experience proves their cost effectiveness.

Many rings equally spaced along the length of the bag, as shown in Fig. 7. 16(a). However, not all manufacturers agree on ring spacing. The argument states that the reduced fabric tension at the bag bottom requires closer ring spacing in order to retain the optimum fabric shape during cleaning, which implies equal fabric deflection in all bag sections as illustrated in Fig. 7. 16(b). Fortune et al., calculate optimum ring spacing to be 2 bag diameters at the bottom and up to a maximum of 4. 5 bag diameters at the top[24].

In order to crack the dust cake on the bags, the values of the flow rate and pressure of the reverse air produced by a blower must be determined for a designer. Because the reverse air bag-house use multi-compartment designs, the cleaning action is one compartment after another. Therefore, the flow rate of the blower is needed just for only one compartment cleaning. For one compartment, the flow rate is 0. 2～0. 3m^3/min per square meter of the filter fabric. For example, if the filter fabric area in one compartment is 1000m^2, the flow rate of the blower for cleaning is 200～300m^3/min. Because the maximum reverse pressure across the bags is about 500Pa to make bag collapse, it is better for the pressure of the blower is at least 500Pa greater than the filtration pressure before cleaning, as shown in Fig. 7. 17.

The high pressure blower can also be replaced by the reverse air pulse which is provided by an

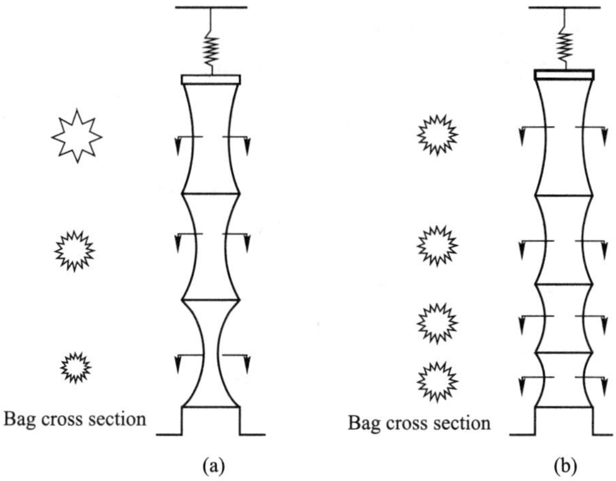

Fig. 7.16 Compensating the increase of bag tension with height by unequal ring spacing
(a) Equal ring spacing; (b) Compensating ring spacing

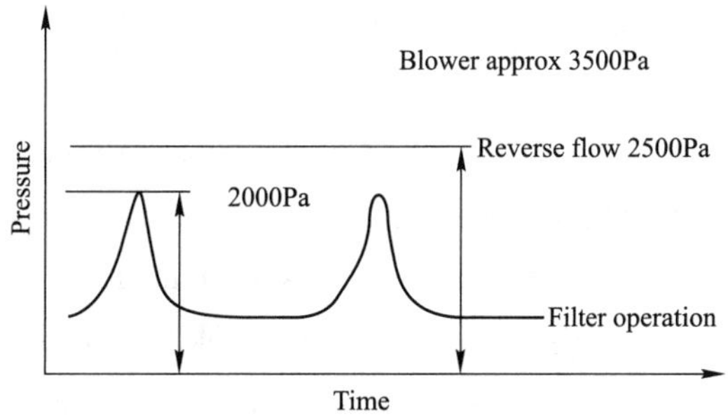

Fig. 7.17 Schematic diagram of the requirements for pressure increase in backwash

air compressor (≥0.3MPa). The cleaning duration of the pulse reverse air flow is tens of seconds (20~50s). This reverse air pulse clean method are usually used in the circle backwash bag-house. In this case, less reverse air flow rate is used because usually less than 20 bags are needed to be cleaned rather than one compartment. The reverse air velocity is about 1.5 ~ 2 times of the filtration velocity. For example, if there are 10 bags with diameter 200mm and length 6000mm under a turning reverse compressed air tube, and the filtration velocity is 0.8 m/min, the flow rate supplied by the compressor is about 45~60m^3/min.

7.6.3.2 Pulse Jet

The term ‘pulse jet cleaning’ initially implied the presence of a venturi. This usage is no longer universal, and the term pulse jet cleaning as used here refers to cleaning action brought about by short time (50 ~ 200ms), high pressure pulses created by any technique, with or without a venturi. A pulse jet cleaning system mainly consists of the jet tubes, compressed air reservoir, and electrical solenoid valve, as shown in Fig. 7.18.

The compressor maintains a reservoir feeding the header pipe at the desired high air pressure. A

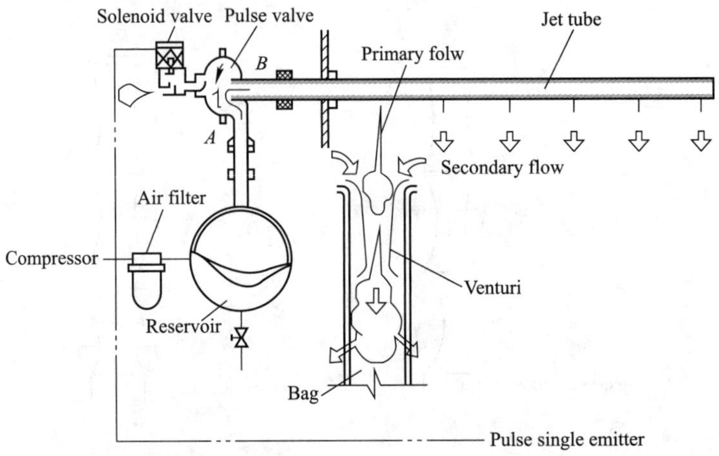

Fig. 7. 18 Schematic diagram of a pulse jet cleaning system

pneumatically operated diaphragm valve, actuated by an electrical solenoid valve, is a typical arrangement for delivering the cleaning pulses to the individual blow pipes and the bag beneath them. Signals for opening and closing the solenoid valve now almost universally originate with solid state switches and timing circuits. These components should last longer than their mechanical predecessors and reduce the number of system failures attributable to this source.

A major distinction between reverse air cleaning and pulse jet cleaning is that the time scale of the pulse jet is a fraction of a second (50~150ms is typical) in duration, while the reverse air cleaned bag-houses described in the previous part have much longer durations (tens of seconds).

The hardware of the pulse jet differ completely from the reverse air. For instance, the most common bag mounting hardware of pulse jet technology is the cage, a cylindrical structure composed of steel wires running parallel to the longitudinal axis of the bag and supported by steel rings at evenly spaced intervals along its length. However, in reverse air system, the anti-collapse rings and tensioning mechanisms are used.

Typically, the pulse passes through a venturi nozzle (at the mouth of each bag) which acts as a jet pump, aspirating a large volume of gas from the exhaust plenum and converting this large volume of gas into high pressure pulse. The pulse becomes a bag-flexing shock wave which traverses the bag length and rebounds from the closed bottom end of the bag.

That pulse propagation occurs as the 'bubble' action shown in Fig. 7. 19(a) is less certain than the universally observed dust removal brought about by the pulse. Other representations of the pulse cleaning action picture the motion as one in which the entire bag expands outward, like a long balloon being blown up [as shown in Fig. 7. 19(b)], as the shock wave travels down the bag[25]. When the pressure inside the bag decreases following valve closure, the bag collapses back on the cage more or less uniformly. These fast expansion and collapse movements shack the dusts off the bags.

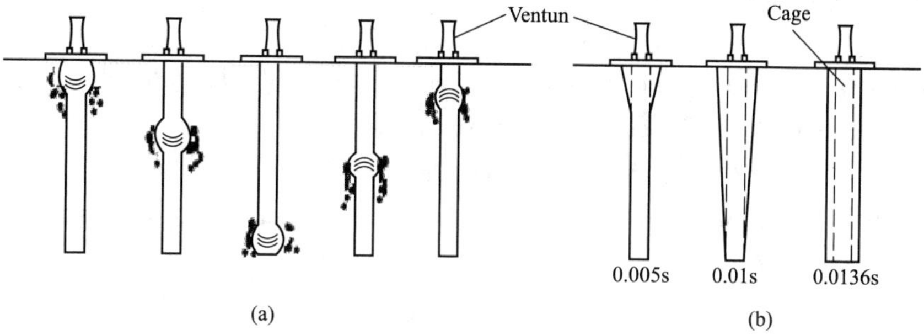

Fig. 7.19 Models of pulse jet bag cleaning mechanism
(a) Bobble model; (b) Balloon model

To determine the air volume of the pulse jet flow theoretically is difficult. Then a empirical calculation method is suggested. Based on the designing requirement, the cleaning velocity should be two times higher than filtration velocity. If the secondary flow is not considered, the cleaning air volume can be obtained from the bag diameter and length. It is given by

$$Q_R = 2v\pi\phi l \tag{7.37}$$

Where Q_R——the volume of one jet in m^3/min;

v——the filtration velocity in m/min;

ϕ——the bag diameter in m;

l——the bag length in m.

If there are m bags beneath the blow tube, the air volume of one pulse valve in one pulse jet cleaning duration t is given by

$$Q = 2mv\pi\phi lt \tag{7.38}$$

Then the total flow rate of the compressed air is (1.5 security coefficient is considered) expressed as

$$Q_T = 1.5nQ/T \tag{7.39}$$

Where Q_T——the total flow rate of the compressed air in m^3/min;

n——the number of the pulse valve;

T——the period of the pulse jet in min, which can be determined from Table 7.6.

Table 7.6 Suggested pulse jet period

Inlet particle concentration/$g \cdot m^{-3}$	$\leqslant 5$	5~10	$\geqslant 10$
Pulse jet period T/min	25~30	20~25	10~20

Example 7.5 Determination of the compressed air flow rate of a pulse jet bag-house. The conditions are given as $n=10$, $m=16$, $v=1m/min$, $c_0=10g/m^3$, $\phi=0.16m$, $l=4m$, and $t=0.1s$.

Solution

The air volume of one pulse valve per time is

$$Q = 2mv\pi\phi lt = 0.107\text{m}^3/\text{time}$$

Because the inlet particle concentration is 10g/m^3, from Table 7.6, the pulse period is 25min. Then, the total flow rate of the compressed air is

$$Q_T = 1.5nQ/T = 0.0642\text{m}^3/\text{min} \approx 65\text{L/min}$$

According to the flow rate of the compressed air, a compressor can be selected.

Exercises

7.1 If the gas velocity of flowing around a fiber is $v_0 = 1.0\text{m/min}$, compute the diffusion, interception, and inertial impaction collection efficiency of a spherical fiber of 40μm in diameter to remove the particles of 1μm, 5μm, and 10μm ($\mu = 1.85\times10^{-5}\text{Pa}\cdot\text{s}$, $\rho = 1.2\text{kg/m}^3$).

7.2 The thickness of the cloth of a bag filter with the solid fraction 0.3 is 30 mm, the fiber of the cloth is 40μm in diameter, and the filtration velocity is 1.2m/min. Calculate the fractional collection efficiency of the bag filter for the particles of 2μm diameter ($\mu = 1.85\times10^{-5}\text{Pa}\cdot\text{s}$, $\rho = 1.2\text{kg/m}^3$).

7.3 Calculate the pressure drop of the clean fabric filter, and the diffusion efficiency for 0.1μm particle at the standard condition. The calculating data includes that filtration velocity is $v_0 = 1.2\text{m/min}$, the filter porosity is $\varepsilon = 0.6$, the filter thickness is $L = 2\text{mm}$, and the fiber diameter is $d_f = 50\mu\text{m}$.

7.4 The thickness of the fabric median is 2mm, the fiber diameter of this fabric median is 30μm, and the filtration velocity is $v_0 = 1.2\text{m/min}$. Plot the pressure drop curve of the solid fraction of the fabric median in the range from 0.1 to 0.6 at standard condition.

7.5 The thickness of the a filter bed is 2mm for use in removing radioactive particles from a gas stream, and the emission concentration is 2mg/m^3. If the emission concentration is needed to be reduced to 1mg/m^3, determine the necessary thickness of the filter.

7.6 A baghouse filter is running at a constant gas rate for 30min. During which period, 100m^3 of gas from a cement kiln operation is processed. The initial and final pressures in the unit are 500Pa and 2000Pa respectively. From equation (7.35), the relation between the pressure and the flow rate can be written as $\Delta P = AQ + BQ^2t$. If the filter is further operated for 1h at the final pressure, calculate the quantity of additional gas treated.

7.7 A baghouse filter is used to treat the fly ash at the filtration velocity of 1.2m/min. When the filter is clean, the pressure is 500Pa. After 20min, the pressure is increased to 1500Pa. It is known that the particle concentration of the gas is 5g/m^3, the particle density is $2\times10^3\ \text{kg/m}^3$, and the dust cake porosity is 0.3. Then, give the equation of the relation between the pressure and the dust cake thickness (hint: $\Delta P = Cv_0 + Dv_0Ht$, where C and D are constant, and H is the dust cake thickness).

7.8 A filter has an overall collection efficiency of 99% when the particle concentration is 5g/m^3, the filtration velocity is 1.5m/min, the thickness of the filter is 3mm, the filter porosity is $\varepsilon = 0.6$, and the dust cake porosity is 0.3. The initial and final pressure in the unit are 500Pa and 1500Pa during 20min. If the pressure is greater than 2500Pa, the particulate on the filter must be cleaned. Determine the dust clean period.

7.9 (1) If $v_0 = 1.5\text{m/min}$, how many square meters of filter surface would be needed for a 750MW coal fire power plant that produce $3\times10^6\ \text{m}^3/\text{h}$ flow rate of stack gas?

(2) If the filter area is in the form of cylindrical bags 10m long and 150mm in diameter, how many of them will be needed?

References

[1] Riefler N, Ulrich M, Morshauser M, et al. Particle penetration in fiber filters [J]. Particuology, 2018, 40: 70-79.

[2] Emi H. Fundamentals of particle separation and air filters [J]. Journal of Aerosol Science, 1991, 22: 722-730.

[3] Kothari V K, Das A. , Singh S. Filtration behaviour of woven and nonwoven fabrics [J]. Indian Journal of Fibre and Textile Research, 2007, 32: 214-220.

[4] Endo Y, Chen D, Pui D Y, et al. Effects of particle polydispersity and shape factor during dust cake loading on air filters [J]. Powder Technology, 1998, 98 (3): 241-249.

[5] Wang C. Electrostatic forces in fibrous filters. A review. [J]. Powder Technology, 2001, 118 (1): 166-170.

[6] Michael J M, Clyde O. *Filtration-Principles and Practices* [M]. New York: Marcel Dekker, Inc. 2017.

[7] Saleem M, Krammer G. Effect of filtration velocity and dust concentration on cake formation and filter operation in a pilot scale jet pulsed bag filter [J]. Journal of Hazardous Materials, 2007, 144 (3): 677-681.

[8] Cao M, Gu F, Rao C, et al. Improving the electrospinning process of fabricating nanofibrous membranes to filter PM2. 5 [J]. Science of the Total Environment, 2019, 666: 1011-1021.

[9] Choi D Y, Jung S, Song D, et al. Al-coated conductive fibrous filter with low pressure drop for efficient electrostatic capture of ultrafine particulate pollutants [J]. ACS Applied Materials & Interfaces, 2017, 9 (19): 16495-16504.

[10] Jegatheesan V, Vigneswaran S. Deep bed filtration: mathematical models and observations [J]. Critical Reviews in Environmental Science and Technology, 2005, 35 (6): 515-569.

[11] Neiva A C, Goldstein L. A procedure for calculating pressure drop during the build-up of dust filter cakes [J]. Chemical Engineering and Processing, 2003, 42 (6): 495-501.

[12] Li M, Feng Y, Wang K Y, et al. Novel hollow fiber air filters for the removal of ultrafine particles in PM2. 5 with repetitive usage capability [J]. Environmental Science & Technology, 2017, 51 (17): 10041-10049.

[13] Seifert O E, Schumacher S C, Hansen A C, et al. Viscoelastic properties of a glass fabric composite at elevated temperatures: experimental and numerical results [J]. Composites Part B-engineering, 2003, 34 (7): 571-586.

[14] Ulbricht M. Advanced functional polymer membranes [J]. Polymer, 2006, 47 (7): 2217-2262.

[15] Donovan R P. *Fabric Filtration for Combustion Sources-Fundamentals and Basic Technology* [M]. New York, Marcel Dekker, Inc. 1985.

[16] Saracco G, Montanaro L. Catalytic ceramic filters for flue gas cleaning. Preparation and characterization [J]. Industrial & Engineering Chemistry Research, 1995, 34 (4): 1471-1479.

[17] Mao N, Liu J, Chang D, et al. Discussion of influencing factors on filtration performances of PTFE membrane filters [J]. Information Sciences, 2015.

[18] Jones A. *Membrane and separation Technology* [M]. Canberra, Australian Government Publishing Service. 1987.

[19] Park B H, Kim S B, Jo Y M, et al. Filtration characteristics of fine particulate matters in a PTFE/Glass

composite bag filter [J]. Aerosol and Air Quality Research, 2012, 12 (5): 1030-1036.

[20] Calle S, Contal P, Thomas D, et al. Evolutions of efficiency and pressure drop of filter media during clogging and cleaning cycles [J]. Powder Technology, 2002, 128 (2): 213-217.

[21] Li X, Chambers A J. Model of dust collection and removal from mechanically shaken filter bags [J]. Filtration & Separation, 1995, 32 (9): 891-895.

[22] Choi H, Kim T. Numerical simulation of ultrasonic generator in dust removing system [J]. Advanced Materials Research, 2012: 1446-1450.

[23] Qian F, Ye Y. Effect of the inlet configuration on the inner airflow distribution uniformity of the bag filter [C]. International Conference on Mechanic Automation and Control Engineering, 2011: 6452-6455.

[24] Fortune O F, Miller R L, Samuel E A. Fabric filter operating experience from several major utility units. Third Symposium on the Transfer and Utilization of Particulate Control Technology, EPA-600/9-82-005a, 1982: 82-93.

[25] Dennis R, Hovis L S. Pulse jet filtration theory——a state of the art assessment, Fourth Symposium on the Transfer and Utilization of Particulate Control Technology, EPA-600/9-84-025a, 1984: 22-36.

8 Scrubbers

Scrubbers employ liquid washing to remove the gaseous or particulate pollutants from gas stream.[1] There are many different types of scrubbers available, such as spray tower[2], centrifugal scrubber[3], orifice scrubber[4], impingement plate tower[5], packed tower[6], venturi scrubber[7], etc. The scrubbers are generally classified into two types: low energy scrubbers and high energy scrubbers. The collection efficiency of scrubbers can be related to the total energy loss. In the most case, the higher the scrubber power per unit volume of gas treated, the higher is the collection efficiency.

Scrubbing is a very effective means of removing small particles from a gas. The drops formed in all scrubbers are generally much larger than the particles to be collected[8]. In most case the scrubbing medium is water, and occasionally a different substance is used. Different types of scrubbing devices employ different means of forming the water droplets and different means of ensuring a relative velocity between the water droplets and the gas to be cleaned. In all cases the cleaning mechanism involves attachment of the particulates to the droplets. The droplets are then collected and drained to a sump. Further processing is then required to remove the collected substance from the water before the water is discharged or reused[9].

In this chapter, the mechanisms of the particle collection by a single spherical droplet will be discussed first. Then the effect of a large number of droplets sweeping across the gas will be considered. Afterward, the collection performances of three typical scrubbers, spray tower, packed tower, and venturi scrubber are going to be analyzed where the results of our study of the collection mechanisms will be applied to the scrubbers.

8.1 Mechanisms of Particle Capture of a Single Droplet

In scrubbers the particles are collected by droplets or by wetted surfaces. Because there are various types of the wetted surface collecting medium, very few research works of discussing the wetted surface collecting mechanism had been found in academic literature. Since the droplets are very regular for theoretical analysis, many researchers have studied the particle collection efficiency of the liquid droplets. So, in this text book, only the mechanisms of the water droplets are discussed. As the same as a fabric fiber, the mechanisms of the water droplets to collect and retain smaller particles, either solid or liquid, are interception, inertial impaction, and diffusion[2].

8.1.1 Interception

We now consider a collection efficiency for a single droplet. This efficiency is defined as the ratio of the number of particles collected to the number of particles initially contained in the volume swept through by the droplets.

Interception takes place when a particle flowing the streamlines of flow a spherical target. Thus, if the streamline on which the particle center lies is within a distance $d_p/2$ of the sphere, interception occurs, as shown in Fig. 8. 1.

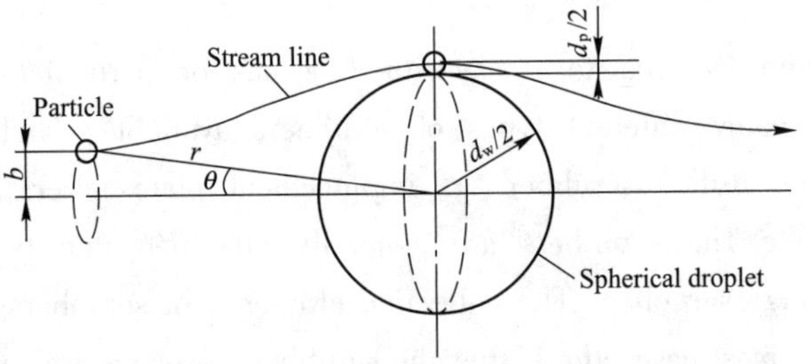

Fig. 8. 1 Deposition of particles on a sphere by interception

Along this stream line, the particle of given diameter d_p will strike the droplet if it lies initially within a certain distance b of the axis of motion of the droplet. It means that all the particles with d_p in the area of πb^2 will be captured by the droplet. Then the interception collection efficiency is expressed as

$$\eta_R = \frac{\pi b^2}{\pi d_w^2/4} = \frac{4b^2}{d_w^2} \tag{8.1}$$

Where d_w——the droplet diameter.

The problem is to determine the distance b from the stream line where the particle located at $r = (d_w + d_p)/2$, and $\theta = \pi/2$.

If the Reynolds number of droplet $Re = \rho v d_w/\mu \gg 1$, the flow field belongs to the potential flow. The stream function is given by

$$\psi = \frac{1}{2} v_0 \left[1 - \left(\frac{d_w}{2r}\right)^3\right] r^2 \sin^2\theta \tag{8.2}$$

Where v_0——the gas velocity far from the droplet.

When $\theta = \pi/2$, $r = (d_w + d_p)/2$, from equation (8. 2), we have

$$\frac{1}{2} v_0 \left[1 - \left(\frac{d_w}{2r}\right)^3\right] r^2 \sin^2\theta = \frac{1}{2} v_0 \left[1 - \left(\frac{d_w}{d_w + d_p}\right)^3\right]\left(\frac{d_w + d_p}{2}\right)^2 \tag{8.3}$$

When $r \to \infty$, in Fig. 8. 1, $r \sin\theta = b$, equation (8. 3) becomes

$$b^2 = \frac{1}{4}\left[(d_w + d_p)^2 - \frac{d_w^3}{d_w + d_p}\right] \tag{8.4}$$

When equation (8. 4) is substituted in equation (8. 1), the interception efficiency in the potential flow is obtained

$$\eta_R = (1 + G)^2 - \frac{1}{1 + G} \tag{8.5}$$

Where G——the interception parameter given by

$$G = d_p/d_w \tag{8.6}$$

If $d_w >> d_p$, then $G = d_p/d_w << 1$, equation (8.5) can be further simplified as

$$\eta_R = 2G = 2d_p/d_w \tag{8.7}$$

8.1.2 Inertial Impaction

The particle will strike directly onto a droplet because the particle is unable to follow the rapidly curving streamlines around the droplet due to the inertia effect. It is called inertial impaction. Therefore, the particle trajectory in the inertial impaction is difficult to be determined because the particle does not follow its streamline. The mechanism analysis indicates that the inertial impaction efficiency is the function of the Stokes number. The Stokes number is given by

$$\mathrm{St} = \frac{\rho_p d_p^2 v_0}{18\mu d_w} \tag{8.8}$$

In many scrubber applications the inertial impaction is the predominant removal mechanism, especially for particle larger than 1μm in diameter. In that case Calvert[10] has suggested an empirical collection efficiency of inertial impaction for a single droplet. It is given by

$$\eta_I = \left(\frac{\mathrm{St}}{\mathrm{St} + 0.35}\right)^2 \tag{8.9}$$

8.1.3 Diffusion

When the particle size is smaller than 0.5μm, the diffusion could become a dominant capture mechanism.

Diffusion collection efficiency is related to the Peclet number, Pe, which is defined as

$$\mathrm{Pe} = \frac{v_0 d_p}{D} \tag{8.10}$$

Where D——the diffusion coefficient given by equation(3.2).

According to equation (3.35), the concentration distribution near the sphere due to diffusion mechanism is expressed as

$$c = c_0 \mathrm{erf}\left(\frac{x}{\sqrt{4Dt}}\right) \tag{8.11}$$

Where x——the distance from the any point in the space to the surface of a drop.

By Fick's first law, the particulate mass deposited on the per square meter of the drop surface is given by

$$F = D\left(\frac{\partial c}{\partial x}\right)_{x=0} = c_0\sqrt{D/\pi t} \tag{8.12}$$

The time of the drop remained in the spray zone is given by

$$t_0 = h/v_0 \tag{8.13}$$

Where h——the height of the spray zone.

During time t_0, the mass deposited on the drop surface is expressed as

$$m = 4\pi d_w^2 \int_0^{t_0} c_0(\sqrt{D/\pi})t^{-1/2}\mathrm{d}t = 8\pi d_w^2 c_0\sqrt{Dt_0/\pi} \tag{8.14}$$

The total mass of the particles flowing toward the drop in its project area $\pi d_w^2/4$ is expressed as

$$m_0 = t_0 c_0 v_0 \frac{\pi}{4} d_w^2 \tag{8.15}$$

Then, the collection efficiency of a single drop is given by

$$\eta_D = \frac{m}{m_0} = \frac{32}{v_0}\left(\frac{D}{\pi t_0}\right)^{1/2} = 32\left(\frac{d_p}{\pi v_0 t_0}\right)^{1/2} \mathrm{Pe}^{-1/2} \tag{8.16}$$

8.1.4 Combination Collection Efficiency of Independent Mechanisms

When airborne particulate exposed in liquid spray, interception, inertial impaction and diffusion effects occur simultaneously. The efficiencies calculated for each effect need to be combined into a single efficiency. The actual combination of effects is quite complex. An acceptable collection efficiency due to combined effect of three independent mechanisms caused by one drop can be approximately given by

$$\eta = 1 - (1 - \eta_R)(1 - \eta_I)(1 - \eta_D) \tag{8.17}$$

If the particle size is larger than 1μm, the diffusion effect can be neglected, and if the particle size is smaller than 1μm, the inertial impaction effect is negligible. It is means that the lowest efficiency of a spray tower appears at the particle diameter of about 1μm.

8.2 Overall Efficiency by Monodisperse Drops

8.2.1 Cylindrical Model

In order to achieve a satisfactory collection efficiency, a large number of drops are required in a scrubber operation. Thus, the particles will be exposed repeatedly to the action of a series drops. Martin had suggested a cylindrical model as follows[10].

Consider a long cylinder of d_w in diameter, where d_w is the diameter of the drops, and examine what happens to the particles which initially enter the cylinder in the opposite direction to the drops, as shown in Fig. 8.2.

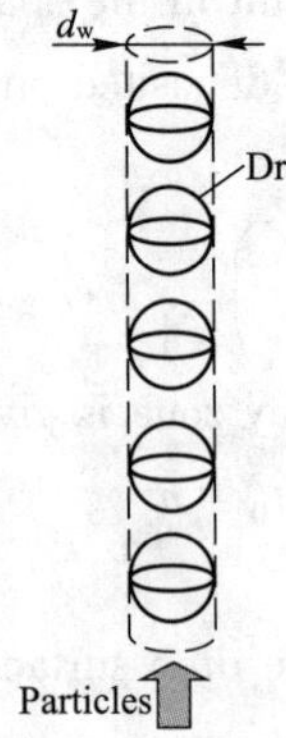

Fig. 8.2 Cylindrical model of overall collection efficiency

Let c_0 be the particle concentration originally entering the cylinder, let c_1 be the particle concentration which pass the first drop, and so on, until c_k particle concentration remains in the cylinder after passing k drops. Then, after first drop, c_1 is expressed as

$$c_1 = c_0(1 - \eta)$$

After second drop, c_2 is expressed as

$$c_2 = c_1(1 - \eta) = c_0 (1 - \eta)^2$$

After k drops, c_k is expressed as

$$c_k = c_0 (1 - \eta)^k$$

The collection efficiency after k drops is expressed as

$$\eta_s = 1 - \frac{c_k}{c_0} = 1 - (1 - \eta)^k \tag{8.18}$$

Because the collection efficiency of one drop η is very small, that is $\eta \ll 1$, then $(1 - \eta)^k$ can be converted to series, given by

$$(1 - \eta)^k = 1 - k\eta + \frac{k(k - 1)}{2!}\eta^2 - \cdots \approx 1 - k\eta \tag{8.19}$$

Then, equation (8.18) becomes

$$\eta_s = k\eta \tag{8.20}$$

In this cylindrical model, the overall collection efficiency caused by a series drops is overestimated by equation (8.18) because the spaces between the drops on the cross section of the gas flow direction are not considered. This may lead to the collection efficiency too great, usually be equal to 100%.

8.2.2 Box Model

Because the cylindrical model is not much reasonable to predict the overall collection efficiency by monodisperse drops, here we will propose a box model to develop the overall partcle collection efficiency equation by the drops with the same size. Suppose the spray zone looks like a box. In this box, the drops with number of $i \times j \times k$ are distributed uniformly, as shown in Fig. 8.3.

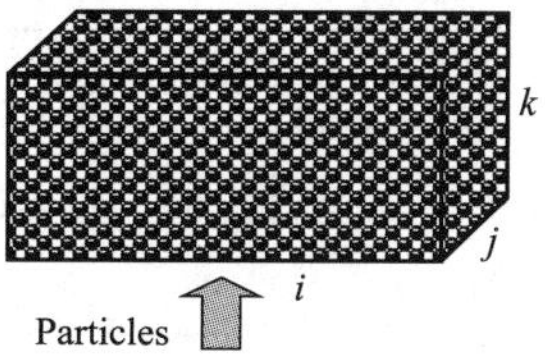

Fig. 8.3 Box model of overall collection efficiency

Let c_0 be the particle concentration originally entering the cylinder with velocity v_0. The mass of the particles collected by the first layer of the drops is given by

$$m_1 = v_0\left(\frac{\pi}{4}d_w^2 ij\right) c_0\eta \tag{8.21}$$

The total mass of the particles entering the box is given by

$$M = Qc_0 \tag{8.22}$$

Then, the collection efficiency of the first drop layer is given by

$$\eta_1 = \frac{m_1}{M} = \frac{\pi v_0 d_w^2}{4Q} ij\eta \tag{8.23}$$

Let

$$n_A = ij \tag{8.24}$$

Where n_A ——the number of the drops in a drop layer.

Then equation (8.23) can be written as

$$\eta_1 = \frac{\pi v_0 d_w^2}{4Q} n_A \eta \tag{8.25}$$

Because in any drop layer, the fractional collection efficiency is equal to each other. For the k layers in series, the overall collection efficiency is given by

$$\eta_s = 1 - (1 - \eta_1)^k \tag{8.26}$$

In equation (8.26), since $\eta_1 \ll 1$, then equation (8.26) becomes

$$\eta_s = k\eta_1 \tag{8.27}$$

Let the concentration of the droplets with diameter d_w be N_w in the spray, the number of the drops per meter is $N_w^{1/3}$. If the spray zone is rectangular, then

$$i = aN_w^{1/3}, \quad j = bN_w^{1/3}, \quad k = hN_w^{1/3} \tag{8.28}$$

Where a, b and h——the length, width and height of the rectangle respectively.

Actually, for any shape of the spray zone, the number of the drops in a drop layer is expressed as

$$n_A = ij = abN_w^{2/3} = AN_w^{2/3} \tag{8.29}$$

Where A——the scrubber section area.

Example 8.1 The operation data of a water spray tower are given in Table 8.1. If the number concentration of 1mm water drops in water spray tower is $1 \times 10^7/\text{m}^3$. Calculate the overall collection efficiency for particle diameter of 1μm based on cylindrical model and box model respectively.

Table 8.1 Operation data of a spray tower

A/m^2	H/m	$v_0/\text{m}\cdot\text{s}^{-1}$	$Q/\text{m}^3\cdot\text{s}^{-1}$	$\rho/\text{kg}\cdot\text{m}^{-3}$	$\mu/\text{Pa}\cdot\text{s}$	$\rho_p/\text{kg}\cdot\text{m}^{-3}$	$D/\text{m}^2\cdot\text{s}^{-1}$
1	2	1	1	1	1.85×10^{-5}	2×10^3	2.7×10^{-11}

Solution

(1) Interception:

$$\eta_R = 2G = 2d_p/d_w = 2 \times 10^{-3}$$

(2) Inertial impaction:

$$\text{St} = \frac{\rho_p d_p^2 v_0}{18\mu d_w} = 0.003$$

$$\eta_I = \left(\frac{\text{St}}{\text{St} + 0.35}\right)^2 = 7 \times 10^{-5}$$

(3) Diffusion:

$$\mathrm{Pe} = \frac{v_0 d_p}{D} = 0.37 \times 10^5$$

$$\eta_D = 32\left(\frac{d_p}{\pi v_0 t_0}\right)^{1/2} \mathrm{Pe}^{-1/2} = 7 \times 10^{-5}$$

(4) Combined collection efficiency:

$$\eta = 1 - (1 - \eta_R)(1 - \eta_I)(1 - \eta_D) = 0.0021$$

(5) Overall collection efficiency:

1) Cylindrical model:

$$k = hN_w^{1/3} = 430$$

$$\eta_s = k\eta = 0.903 = 90.3\%$$

2) Box model:

$$n_A = AN_w^{2/3} = 46418$$

$$\eta_1 = \frac{\pi v_0 d_w^2}{4Q} n_A \eta = 0.77 \times 10^{-4}$$

$$\eta_s = k\eta_1 = 0.033 = 3.3\%$$

From this example we can see that the collection efficiency (90.3%) calculated by cylindrical model is too overestimated. Thus, the box model could be much more reasonable. In fact, it is difficult to remove 1μm particle efficiently from the flume by using the spray tower if the water drop diameter is larger than 1mm, which will be discussed in the next section.

The relation of the water to gas ratio and the number of the drops per cubic meter is expressed as

$$\gamma_w = \frac{\pi}{6} d_w^3 N_w \tag{8.30}$$

In this example, water to gas ratio is quite high. The water to gas ratio of number concentration 10^7 with drop diameter of 1nm is 5.2L(water)/m^3 (gas). In a industrial spray tower, the water to gas ratio is usually ranged from 0.5~2L/m^3.

8.3 Overall Efficiency by Drops with Size Distribution

The overall collection efficiency ought to be found for industrial application of scrubbers. However, in the real cleaning process of scrubbing, the droplet size and particle size are not uniform. Because the collection efficiency is the function of water droplet diameter, particle diameter, and ratio of water to gas flow rate, calculating the overall collection is very difficult in this case.

Fig. 8.4 is the collection efficiency of a spray chamber as a function of droplet diameter and particle diameter given by Richard, and Seinfeld[11]. It is clearly shown that when the diameter of the droplets is smaller than 1000μm and the particle diameter is larger than 5μm, the collection efficiency approaches 100%. However, for 1μm particles, the collection efficiency is only about

0.25% when drop diameter is 1000μm. If we want to achieve high collection efficiency for 1μm particles, the drop diameters should be in the ranges of 50~100μm.

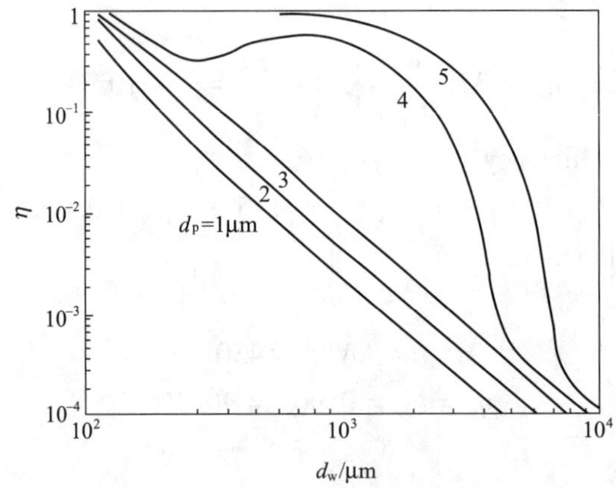

Fig. 8.4　Collection efficiency of a spray tower as a function of droplet and particle diameter

(Particle density of 10×10^3 kg/m^3, chamber diameter and height are 1 and 5m, and water to gas ratio is 1L/m^3)

It is difficult to calculate the overall collection efficiency because the particles and water droplets vary in size in a spray tower. Thus, we have to find a simplified method to estimate the overall collection efficiency of water spray tower. Zhao and Pfeffer[12] found that the overall collection efficiency of electrostatic precipitators could be predicted approximately when the particle diameter is substituted by the median particle diameter particle. Therefore, if both drop diameter d_w and the particle diameter d_p are replaced by median drop diameter d_{w50} and median particle diameter $d_{p_{50}}$ respectively in the box model, the overall collection efficiency of a water spray tower can be predicted. The calculation processes are given as following example.

Example 8.2　Calculate the overall collection efficiency when particle and water drop distribution is considered. Assume the particle size distribution in a particulate pollutant gas and the drop size distribution in spray follow the log-normal distribution. The median particle diameter is 1μm, the median drop diameter is 100μm, and the water to gas ratio is 1L/m^3. Other operation data are given in Table 8.1.

Solution

(1) Interception:

$$\eta_R = 2G = 2d_{p_{50}}/d_{w_{50}} = 2\times10^{-2}$$

(2) Inertial impaction:

$$\mathrm{St} = \frac{\rho_p d_{p_{50}}^2 v_0}{18\mu d_{w_{50}}} = 0.03$$

$$\eta_I = \left(\frac{\mathrm{St}}{\mathrm{St}+0.35}\right)^2 = 6.2\times10^{-3}$$

(3) Diffusion:

$$\mathrm{Pe} = \frac{v_0 d_{p_{50}}}{D} = 0.37\times10^5$$

$$\eta_D = 32\left(\frac{d_{p_{50}}}{\pi v_0 t_0}\right)^{1/2} Pe^{-1/2} = 7 \times 10^{-5}$$

(4) Combined collection efficiency:

$$\eta = 1 - (1 - \eta_R)(1 - \eta_I)(1 - \eta_D) = 0.0262$$

(5) Overall collection efficiency by box model for both particles and drops having size distribution

$$N_w = \frac{6\gamma_w}{\pi (d_{w_{50}})^3} = 1.9 \times 10^9 m^{-3}$$

$$k = hN_w^{1/3} = 2.48 \times 10^3$$

$$n_A = AN_w^{2/3} = 1.54 \times 10^6$$

$$\eta_1 = \frac{\pi v_0 d_{w_{50}}^2}{4Q} n_A \eta = 3.17 \times 10^{-4}$$

$$\eta_s = k\eta_1 = 0.786 = 78.6\%$$

This result indicates when the median diameter of drops is about 100μm, the higher collection efficiency can be obtained in a spray tower. Because the particles of about 1μm are most difficult to be collected, it can be deduced that if the particle median diameter in a pollutant gas is larger or smaller than 1μm, and the overall collection will be higher than 78.6%.

The key point is to generate the drops around 100μm. Fortunately, today many kinds of nozzles can be selected to produce fine water droplets. Of course, the droplets emitted from a nozzle could not be too small. If the median diameter of the drops is smaller than 50μm, the fine drops may be brought by the gas flow out of the scrubber.

8.4 Performance of Scrubbers

8.4.1 Spray Tower

Spray tower, sometimes called spray chamber, is a low energy scrubber. It is perhaps one of the simplest devices of removing the particulate pollutant from an airstream by means of scrubbing method. Fig. 8.5 shows a cross-sectional view of a spray tower. The water are introduced at the top of an empty chamber. The water droplets are sprayed from a series atomizing nozzles while the dirty gas enters the bottom of the chamber and flow upward, encountering the drops which fall to the bottom by gravity. The particles are captured by the drops. The particle-containing drops are then collected in a pool at the bottom to form the slurry. The slurry must be pumped out for treatment to remove the solids, and the cleaned water is usually recycled to the tower.

The performances of a spray tower include the collection efficiency, the gas velocity in chamber, the water to gas ratio, the drop size, and the pressure drop.

8.4.1.1 Collection Efficiency

For the particles of $PM_{2.5}$, the collection efficiency is less than 50% in the most cases. It is

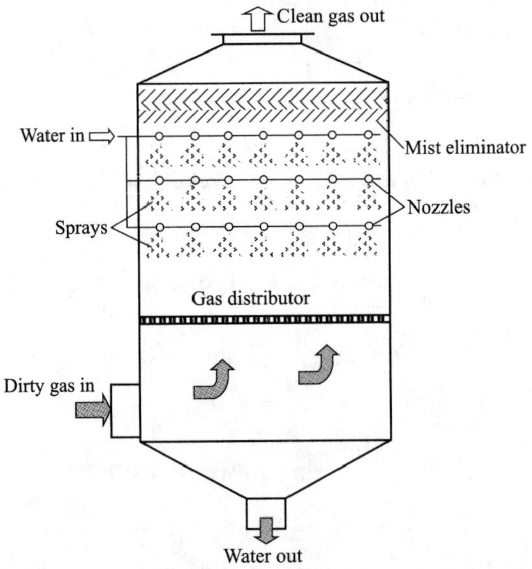

Fig. 8. 5 Spray tower

important to spray drops with uniform size about 100μm, and to distribute the spray in the scrubbing zone evenly in spray tower operation for maintaining a good collection performance.

We have known that the efficiency is relatively low. However, the spray tower is the cheapest and simplest of scrubber. The collection mechanism of the spray tower is a fundamental principle of other scrubbers. The spray scrubbing can also be applied into some other kinds of scrubbers, such as packed tower, impingement scrubber, cyclone spray tower, and so on.

8. 4. 1. 2 Gas Velocity and Water to Gas Ratio

In order to prevent the droplets from being carried out of the spray tower, the gas velocity must not be too high. Otherwise, if the velocity is too low, the inertial impaction effect is reduced which leads to lower collection efficiency.

The gas velocity can be theoretically determined by the terminal velocity of a drop for free fall under the influence of gravity from equation (2. 23). The terminal velocities of water drops falling in standard air is given in Table 8. 2.

Table 8. 2 Terminal velocities of water drops falling in standard air

d_w/μm	10	50	100	200	300	400	500	1000	2000
v_0/m · s^{-1}	3×10^{-3}	0. 07	0. 26	0. 68	1. 02	1. 51	2. 04	4. 33	7. 06

As mentioned before, to get higher collection efficiency, the water median diameter is about 100μm. From the Table 8. 2, the terminal velocity of the 100μm drop is only 0. 26m/s. However, in the applications, the designed gas velocity in a spray chamber is usually about 1m/s. Thus, the terminal velocity of the 100μm drop is too low to satisfy the operation condition. The best way to solve this problem is to increase the jet velocity of the spray nozzle and the water to gas ratio. Here, the drop ejecting speed in the range of 10 to 20m/s and the water to gas ratio in the range from 1 L/m^3 to 1.5L/m^3 are suggested.

Example 8.3 Assume the drop diameter is 100μm, the initial velocity of drop ejected from a nozzle is 20m/s, and the gas velocity is 1m/s. How about the distance of the 100μm drop moving downward in the spray chamber?

Solution

According to equation (2.19), v_w is

$$v_w = (-v_0 + \tau g) + (v_{w_0} + v_0 - \tau g)e^{-t/\tau}$$

Where v_w——the drop velocity;

v_{w_0}——the initial velocity of the drop;

v_0——the gas velocity.

The relaxation time is

$$\tau = \frac{\rho_p d_p^2}{18\mu} = \frac{1 \times 10^3 \times (100 \times 10^{-6})^2}{18 \times 1.85 \times 10^{-5}} = 0.03\text{s}$$

If the drop get to the longest distance, $v_w = 0$. Then the time of drop motion is

$$t = -\tau \ln \frac{v_0 - \tau g}{v_{w_0} + v_0 - \tau g} = 0.1\text{s}$$

From equation (2.21), the distance of the 100μm drop moving downward is

$$z = (\tau g - v_0)t + \tau(v_{w_0} + v_0 - \tau g)(1 - e^{-t/\tau}) = 0.7\text{m}$$

The result shows that if the initial velocity of drop ejected from a nozzle is 20m/s, the 100μm drop can go down 0.7m. This may be not good enough for particle scrubbing. Thus, the multistage sprays can be arranged, as shown in Fig. 8.4. The distance between two stages is 0.7m or less.

Higher ratio of water to gas is good for promoting the particle removal effect. The outlet gases must be saturated in scrubber operation. Otherwise, some particles entering drops of water will be re-evaporated, freeing the particles back into the airstream. If the water quantity is sufficient, the fine particle-containing drops may be condensed into a bigger drop instead of being evaporated. Larger dirty drops can fall easily to the bottom of the chamber.

8.4.1.3 Pressure Drop

Without water drops, the pressure loss caused by the gas flow in a spray tower is given as Δp_0. The presence of the drops will increase the pressure drop across the unit. The total pressure drop of the spray tower becomes

$$\Delta p_T = \Delta p_0 + \Delta p_w \tag{8.31}$$

Where Δp_w——the pressure increment created by the water spray in a spray tower.

Assume that the force of the gas flow acting on a droplet is equal to the gravitational force of this droplet. Thus, It is given by

$$F = \frac{\pi}{6} d_w^3 \rho_w g \tag{8.32}$$

The drop number concentration is determined by equation (8.30). That is

$$N_w = \frac{6\gamma_w}{\pi d_w^3} \tag{8.33}$$

Then, the total number of the droplets per square meter is expressed as

$$N = hN_w \tag{8.34}$$

The pressure drop caused by droplets in the chamber is given by

$$\Delta p_w = FN = \frac{\pi}{6} d_w^3 \rho_w g h N_w = \rho_w g h \gamma_w \tag{8.35}$$

Example 8.4 The pressure drop is 300Pa when the dry gas flows in a spray tower. Use example 8.2 to calculate the pressure drop of the spray tower.

Solution

The pressure drop caused by droplets in the chamber is

$$\Delta p_w = \rho_w g h \gamma_w = 1 \times 10^3 \times 9.8 \times 2 \times 1 \times 10^{-3} = 19.6(\text{Pa})$$

The total pressure drop of the spray tower is

$$\Delta p_T = \Delta p_0 + \Delta p_w = 319.6(\text{Pa})$$

8.4.2 Packed Tower

Packed tower is also one of the commonly used scrubbers in removing particulate. It belongs to the low energy scrubber, filled with some form of packing and sprays which supply water or other liquid to flow counter-current to the gas flow. This type of unit supplies intimate liquid contact with the gas stream. Packed tower is also an inexpensive means of absorbing noxious gases into water or other liquids. It is often constructed of fiberglass to resisted corrosion.

Fig. 8.6 illustrates the configuration of a packed tower. Because the wet surface area of the packing medium in a packing tower is great, the collection efficiency is higher than that of the spray tower. And at the same time, the pressure drop is much higher than that of the spray tower either.

In the common packed tower, only one packed bed with identical size of filling is installed. If two different sizes of packing material are used in packed bed as shown in Fig. 8.6, the operation performances can be improved greatly. There are three main advantages:

(1) It is not easy to be plugged up if there is much particulate matter involved when large size packing material is used at the lower part of the chamber in the packed tower.

(2) Since about 90% of the larger particles are collected in first packed bed, for low concentration of the fine particles, the collection effect of packed tower can be enhanced when the second bed with small filling material is installed in the upper position of the packed tower.

(3) The pressure drop can be reduced because the first packed bed with larger size filling has less resistance to gas flow.

The fine drops (midien diamter $d_w \leqslant 0.1$mm) of the first spray should be used to enhance the diffusion effect when fine drops is mixing with the fine particles in gas. The larger drops (midien diamter $d_w \geqslant 1$mm) of the second spray should be used for washing the second packed bed and for reducing the mist emission.

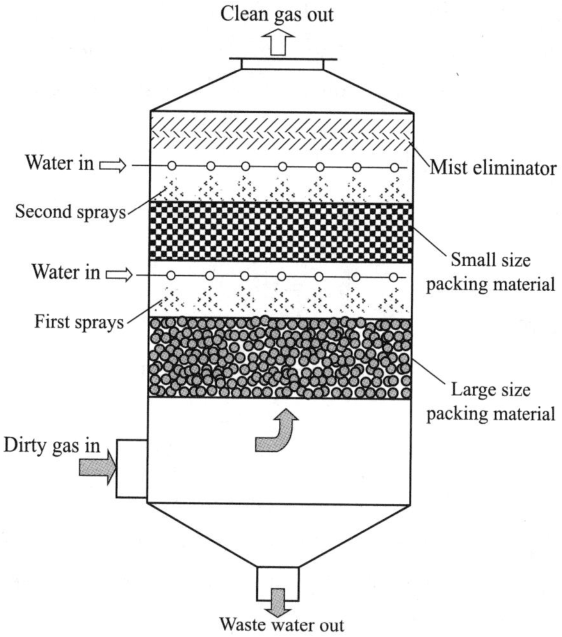

Fig. 8.6 Packed tower with packed bed in series

The packing materials are often made from plastic or ceramic. There are many shapes of the packing materials can be selected, such as the hollow ball, Raschig ring, Lessing ring, Pall ring, saddle ring, etc. The Raschig ring and Lessing ring are more commonly used due to their higher strength and better plugging prevention property, as shown in Fig. 8.7. For particulate separation, the diameters of the packing materials in a packed tower are ranged from 20mm to 80mm.

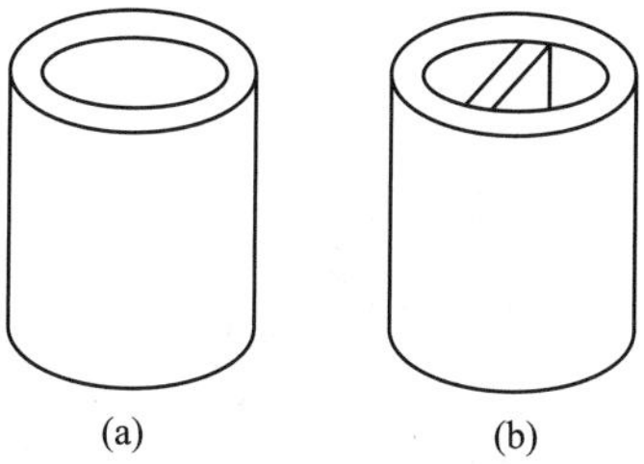

Fig. 8.7 Fillings commonly used in packed tower
(a) Raschig ring; (b) Lessing ring

8.4.2.1 Collection Efficiency

The collection mechanisms of the packed tower are also the interception, inertial impaction, and diffusion. For the packed bed, the interception effect can be negligible because the collection targets are too big. It is difficult to analyze the inertial impaction effect theoretically due to the irregular shape of the fillings and the complicated gas flow field in packed tower. We have known that if the particle is smaller than 1μm, the inertial impaction effect can be neglected. Here we just discuss the diffusion mechanism based on box model.

For any absorbing surface, the particulate mass deposited on the per square meter is given by equation (8.12). Suppose the filling diameter is d_f, the total surface area of fillings in the first packed layer with thickness of d_f is S_1, the time of the dirty gas passing the first layer is t_1, and the particle concentration is c_0. The mass collected on the first layer of the packed bed is given by

$$m_1 = S_1\int_0^{t_1} c_0(\sqrt{D/\pi})t^{-1/2}\mathrm{d}t = 2S_1 c_0\sqrt{Dt_1/\pi} \qquad (8.36)$$

During time t_1, the total mass of the particles getting into the first layer of the packed bed is given by

$$m_0 = t_1 c_0 Q \qquad (8.37)$$

Where Q ——the gas flow rate.

Then, the collection efficiency of the first layer of the packed bed due to the diffusion effect is expressed as

$$\eta_{D_1} = \frac{m_1}{m_0} = \frac{2S_1}{Q}\left(\frac{D}{\pi t_1}\right)^{1/2} \qquad (8.38)$$

For any layer in the packed bed, the collection efficiency is equal to each other. Then the overall collection efficiency of k layers of the packed bed in series is given by

$$\eta_D = 1 - (1 - \eta_{D_1})^k \approx k\eta_{D_1} \qquad (8.39)$$

For any fillings, if the size and shape is selected, the surface area of fillings in a single packed layer S_1 can be determined. Then the overall collection efficiency of the diffusion mechanism effect can be calculated.

Example 8.5 The hollow balls with diameter of $d_f = 30$mm are used in a packed tower. The packed bed is $h=4$m in height and $d=4$m in diameter. The gas velocity in the tower is $v_0=0.5$m/s. Predict the diffusion efficiency for 0.1μm particles ($D=6.8\times10^{-10}$m^2/s).

Solution

The volume of the packed bed is

$$V = Ah = \frac{\pi}{4}d^2h = 50.24(\mathrm{m}^3)$$

The total number of hollow balls in the packed bed is

$$N = V/d_f^3 = 1.86 \times 10^6$$

The layers of the packed bed are

$$k = h/d_f = 133$$

The ball number of one layer is

$$N_1 = N/k = 13984$$

The surface area of the hollow balls in one layer is

$$S_1 = 13984 \times 4\pi d_f^2 = 158(\mathrm{m}^2)$$

The porosity of the packed bed is

$$\varepsilon = 1 - \frac{V_f}{V} = 1 - \frac{\pi d_f^3 N}{6V} = 0.48$$

The gas velocity inside the packed bed is

$$v = v_0/\varepsilon = 1.04(\mathrm{m/s})$$

The time of the gas passing one layer is

$$t_1 = d_f/v = 0.029(\mathrm{s})$$

The flow rate of the tower is

$$Q = \frac{\pi}{4}d^2 v_0 = 6.28(\mathrm{m^3/s})$$

The collection efficiency of one layer is

$$\eta_{D_1} = \frac{2S_1}{Q}\left(\frac{D}{\pi t_1}\right)^{1/2} = 4.35 \times 10^{-3}$$

The overall collection efficiency is

$$\eta_D = k\eta_{D_1} = 0.578 = 57.8\%$$

This result shows that the packed tower has a good collection performance to control the fine particles.

8.4.2.2 Gas Velocity and Water to Gas Ratio

The ranges of the gas velocity and water to gas ratio are quite flexible in the packed towers. The gas velocities are usually ranged from 0.5m/s to 1.5m/s. The water to gas ratios are ranged from 0.3L/m^3 to 1.0L/m^3. Only for the packed bed cleaning, higher water to gas ratio may be used for a very short time interval.

8.4.2.3 Pressure Drop

The pressure drop of the packed tower is higher than that of the spray tower. Generally, the pressure drops of the packed tower are ranged from 1500Pa to 2500Pa. It is difficult to estimate the pressure drop of a packed tower because the size and shape of the fillings are different. For the commercial packing materials, in the most cases, the pressure drop per meter in thickness has been already given by the company. For example, for the packed bed of ϕ40 Raschig ring, when the gas velocity is 1.0m/s, the pressure drop is about 480Pa/m.

8.4.3 Venturi Scrubber

Venturi scrubber is a high energy scrubber which are employed when high collection efficiencies are required and when most of the particles are smaller than 2μm in diameter. A venturi scrubber can only be applied if the particles to be removed are sticky, flammable, or the gas is highly corrosive, or the gas temperature is very high, or for some reasons, other high collection efficiency separators, such as ESPs and fabric filters, cannot be used.

A typical venturi scrubber is shown in Fig. 8.8. The gas flow through a venturi tube which has a throat where the gas is forced to flow at high velocity. Water drops are injected into the air stream near the entrance of the throat to the maximum-velocity section. Gas velocity in the range from 60m/s to 120m/s are achieved, and the high relative velocity between the particle-laden gas flow and the drops promotes collection. When the gas flows from the cylindrical throat to a conical expander where the gas is slowed down to improve the agglomeration between the particles and droplets by means of diffusion. Then the gas carries the dirty drops into a liquid separator, usually a cyclone collector, to collect the drops.

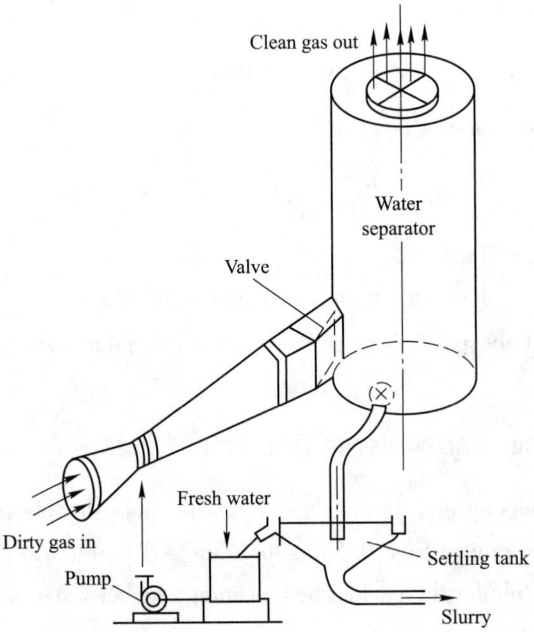

Fig. 8. 8　Venturi scrubber

8. 4. 3. 1　Collection Efficiency

The collection efficiency of a scrubber is quite high. The total collection efficiency of a venturi scrubber can be attributed to three effects including inertial impaction in the throat tube, diffusion in the conical expander, and finally the collection of the water separator. However, no reasonable method has been given for predicting the total collection efficiency of a venturi scrubber till now because the collection effects of the throat tube and the conical expander are very complex.

It is confirmed that the particles captured by the water droplets are more than other kinds of scrubber. If all the particles become the particle-laden water drops, then, the overall collection efficiency of the venturi scrubber will depend on the overall collection efficiency of the water drop separator. Thus, in order to get desirable overall collection efficiency, the design of venturi tube is important.

8. 4. 3. 2　Gas velocity and water to gas ratio

The gas velocity in the throat is crucial in throat design. Fig. 8. 9 shows a cross sectional view of a typical venturi tube.

Generally, gas velocities in the range from 60m/s to 120m/s are selected. Thus, the throat diameter can be determined as

$$D_r = \left(\frac{4Q}{\pi v_i}\right)^{1/2} \tag{8.40}$$

Where D_r——the throat diameter in m;

Q——the flow rate in m^3/s;

v_i——the gas velocity in throat.

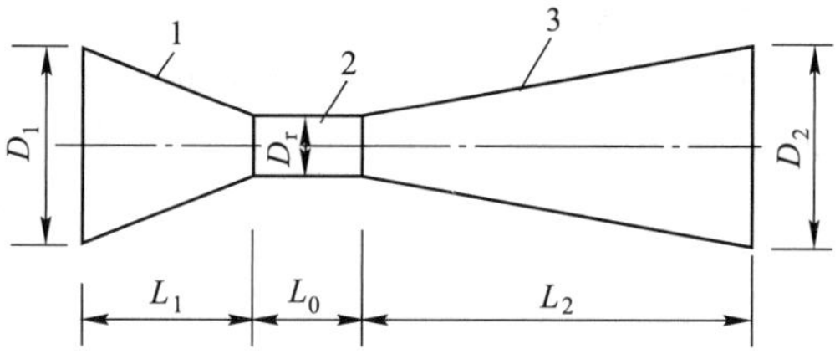

Fig. 8.9 Cross section of a venturi tube

The throat length should be sufficiently long that most of the collection occurs in the throat. If the throat is too short, few particles will be collected in the throat. It would be preferable to use a larger diameter, but longer throat, which would achieve the same collection efficiency with less pressure drop. However, if the throat is too long, larger pressure drop will be caused. The length of throat is suggested as

$$L_0 = 1.5D_r \tag{8.41}$$

The converging conical inlet section diameter of is expressed as

$$D_1 = 2D_r \tag{8.42}$$

The converging conical tube length is expressed as

$$L_1 = \frac{1}{2}(D_1 - D_r)\cot\alpha_1 \tag{8.43}$$

Where α_1——a half of converging conical angle, $\alpha_1 = 12°$.

The expanding conical outlet section diameter is given by

$$D_2 = D_1 \tag{8.44}$$

The expanding conical tube length is given by

$$L_2 = \frac{1}{2}(D_2 - D_r)\cot\alpha_2 \tag{8.45}$$

Where α_2——a half of expanding conical angle, $\alpha_2 = 4°$.

A typical range of water to gas ratios for a venturi scrubber is $1 \sim 3L/m^3$. It is better to use fine spray nozzle. The smaller the droplets, the higher is the collection efficiency.

8.4.3.3 Pressure drop

The venturi scrubbers are operated at pressure drops ranging usually from 5kPa to 20kPa. It is clear that the pressure drop of a venturi scrubber is very high. Thus, we should not employ the venturi scrubbers unless the particulate pollutant control conditions are not permitted.

It is very complex to predict the pressure drop of a venturi scrubber. An emperical equation of the pressure drop of a venturi scrubber suggested by Hesketh[13] is given by

$$\Delta p = \frac{v_i^2 \rho A_i^{0.133} \gamma_w^{0.78}}{1.16} \tag{8.46}$$

Where v_i ——the gas velocity in throat;

ρ ——the gas density;

A_i ——the section area of throat;

γ_w ——the water to gas ratio in L/m^3.

The pressure drop of the water separator is not considered in equation (8. 46).

Example 8. 6 Assume that the section area of the throat is $0.1m^2$, the gas velocity in the throat is 100m/s, the gas density is $1kg/m^3$, and the water to gas ratio is $1.0L/m^3$. Predict the pressure drop of the venturi scrubber.

Solution

The pressure drop of the venturi scrubber is

$$\Delta p = \frac{v_i^2 \rho A_i^{0.133} \gamma_w^{0.78}}{1.16} = 6347\text{Pa}$$

Exercises

8. 1 Calculate the inertial impaction efficiency for particles of 10μm in diameter and with a density of $2.5\times10^3 kg/m^3$ in a gas stream at standard conditions. The aerosol is flowing past a 300μm diameter water drop at a velocity of 2m/s.

8. 2 Calculate the diffusion efficiency for particles of 0. 1μm in diameter in a gas stream at a velocity of 2m/s in 6m high spray tower at standard conditions. The water drop diameter in the spray tower is 300μm.

8. 3 The operation data of a spray tower is as the same as Table 8. 1. Assume that the number concentration of 1mm water drops in water spray tower is $1\times10^7/m^3$. Plot the overall collection efficiency for particle diameter in the range between 1 ~ 10μm based on box model.

8. 4 Calculate the overall collection efficiency when particle and water drop distribution is considered. Assume the particle size distribution in a particulate pollutant gas and the drop size distribution in spray follow the log-normal distribution. The median particle diameter is 5μm, the median drop diameter is 500μm, and the water to gas ratio is $1L/m^3$. Other operation data are given in Table 8. 1.

8. 5 When a packed tower has one packed bed at an surface area of $10000m^2$, the collection efficiency of the packed tower for 0. 5μm particles is 50%. Now, this packed tower will be transformed into two packed bed in series as shown in Fig. 8. 5 to increase the collection efficiency to 75%. How many square meters of the packed bed surface are needed?

8. 6 The Raschig rings of 40mm in diameter and 50mm in height are used in a packed tower. The packed bed is 3m in height and 3m in diameter. The gas velocity in the tower is 1m/s. Predict the diffusion efficiency for 0. 1μm particles ($D=6.8\times10^{-10}m^2/s$).

8. 7 A venturi scrubber is used to treat a particulate pollutant gas stream at a flow rate of $3.6\times10^4 m^3/h$ at standard conditions. If the gas velocity of 80m/s in throat is selected, design the venturi tube and predict the pressure drop of the venturi tube. Assume that the water to gas ratio is $1.0L/m^3$.

References

[1] Noel de Nevers. *Air Pollution Engineering* [M]. New York: McGraw-Hill, 2000.

[2] Darake S, Hatamipour M S, Rahimi A, et al. SO_2 removal by seawater in a spray tower: Experimental study and mathematical modeling [J]. Chemical Engineering Research & Design, 2016: 180-189.

[3] Ali H, Plaza F, Mann A, et al. Flow visualization and modelling of scrubbing liquid flow patterns inside a centrifugal wet scrubber for improved design [J]. Chemical Engineering Science, 2017, 173: 98-109.

[4] Mohebbi A, Taheri M, Fathikaljahi J, et al. Simulation of an orifice scrubber performance based on Eulerian/Lagrangian method [J]. Journal of Hazardous Materials, 2003, 100 (1): 13-25.

[5] Chien C, Tsai C, Sheu S, et al. High-efficiency parallel-plate wet scrubber (PPWS) for soluble gas removal [J]. Separation and Purification Technology, 2015, 142: 189-195.

[6] Ruivo R, Paiva A, Mota J P, et al. Dynamic model of a countercurrent packed column operating at high pressure conditions [J]. Journal of Supercritical Fluids, 2004, 32 (1): 183-192.

[7] Leith D, Cooper D W. Venturi scrubber optimization [J]. Atmospheric Environment, 1980, 14 (6): 657-664.

[8] Silva A M, Teixeira J C, Teixeira S F, et al. Experiments in large scale venturi scrubber Part Ⅱ. Droplet size [J]. Chemical Engineering and Processing, 2009, 48 (1): 424-431.

[9] Chou C, Huang C, Shang N, et al. Treatment of local scrubber wastewater for semiconductor by using photo-catalytic ozonation [J]. Water Science and Technology, 2009, 59 (11): 2281-2286.

[10] Calvert S, Englund H M. *Handbook of Air pollution Technology* [M]. New York: Eds. , Wiley, 1984.

[11] Richard C F, Seinfeld J H. *Fundamentals of Air Pollution Engineering* [M]. New Jersey: Prentice Hall, 1987.

[12] Zhao Z, Pfeffer R. A simplified model to predict the total efficiency of gravity settlers and cyclones [J]. Powder Technology, 1997, 90 (3): 273-280.

[13] Goncalves J A, Alonso D F, Costa M A, et al. Evaluation of the models available for the prediction of pressure drop in venturi scrubbers. [J]. Journal of Hazardous Materials, 2001, 81 (1): 123-140.

9 Hybrid Separators

We have discussed the aerodynamic separation, electrostatic precipitation, fabric filtration, and gas scrubbing in previous chapters. However, as the emission standards getting stricter and stricter, the emission standards sometimes could not be satisfied when the conventional particulate separators were used. If we combined different particle removal means together, some new hybrid separators will be developed. Then the higher collection performances can possibly be achieved.

Today, the development of the hybrid separation technology has become a trend of aerosol particulate control[1-3]. It has been noticed that, in the most cases, the electrostatic force has played an important role in the hybrid separators.

9.1 Aerodynamic Electrostatic Precipitators

9.1.1 Transverse Plate ESP

The aerodynamic separation effect can be made use of to enhance the collection performances. One of the aerodynamic electrostatic precipitators is the transverse plate ESP. As early as 1976, Masuda studied the transverse plate ESP[4]. Later, many kinds of the transverse plate ESP had been developed, such as double fluted plate[5], C plate[6], and double C plate (Yi Chengwu, 2007)[7]. Some of the transverse plate ESPs have been used in kilns, cement factories, and crushers. The typical transverse plate ESP is shown in Fig. 9.1.

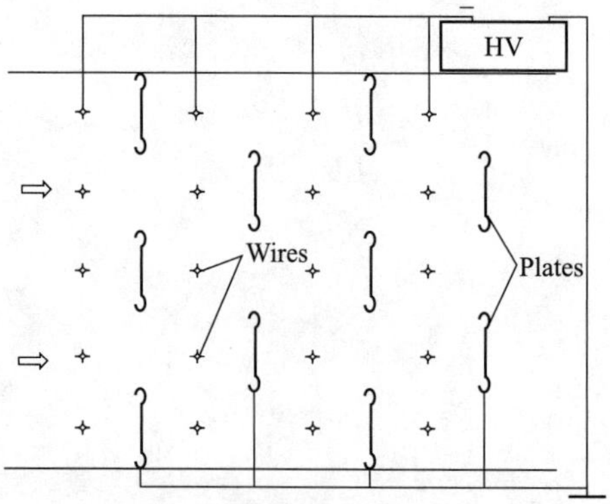

Fig. 9.1 Transverse plate ESP

The flow pattern in the transverse plate ESP is shown in Fig. 9.2. Since the flow is symmetrical to the y axis, the velocity distributions of the first quadrant and the fourth quadrant are needed to discuss only.

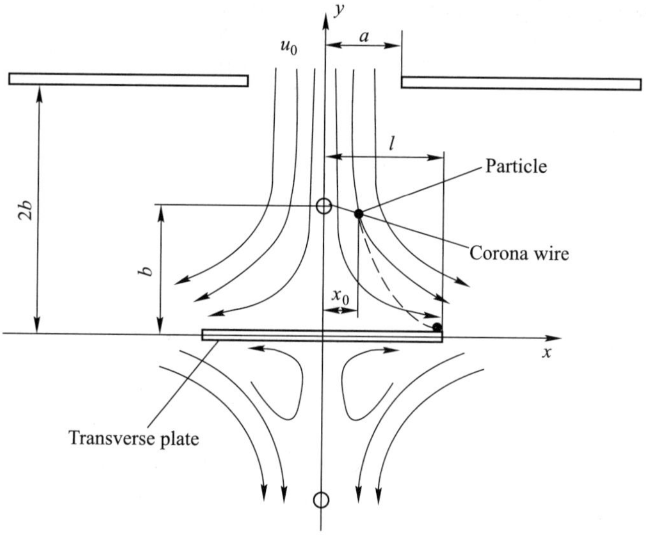

Fig. 9.2 Schematic diagram of flow field around the plate

For the perfect flow, in the front of the transverse plate, the stream function in the first quadrant is[8] given by

$$\psi(x, y) = -Axy \tag{9.1}$$

Where A——a constant.

The velocity distribution in the front of the transverse plate is given by

$$u_x = \frac{\partial\psi}{\partial y} = -Ax, \quad u_y = -\frac{\partial\psi}{\partial x} = Ay \tag{9.2}$$

Suppose that the gas velocity u_0 at $y = 2b$ is approximately equal to the gas velocity at $y = b$, where u_0 is called the average gas velocity in the flow section in width of a, as shown in Fig. 9.2. Then the constant A can be found from equation (9.2), expressed as

$$A = -\frac{u_0}{b} \tag{9.3}$$

Then equation (9.2) becomes

$$u_x = \frac{u_0}{b}x, \quad u_y = -\frac{u_0}{b}y \tag{9.4}$$

Assume that the flow is the laminar flow. The motion equations of the particle in the electric field are expressed as

$$-3\pi\mu d_p(\omega_x - u_x) = m\frac{d\omega_x}{dt} \tag{9.5}$$

$$Eq - 3\pi\mu d_p(\omega_y - u_y) = m\frac{d\omega_y}{dt} \tag{9.6}$$

Where m——the mass of the particle;

ω_x, ω_y——the electrical migration velocity components of a particle.

Substituting $m = \pi d_p^3\rho_p/6$ and equation (9.4) into above motion equations, while the relaxation time $\tau = \rho_p d_p^2 C_c / 18\mu$ is introduced, we have

$$x'' + \frac{1}{\tau}x' - \frac{u_0}{\tau b}x = 0 \tag{9.7}$$

$$y'' + \frac{1}{\tau}y' + \frac{u_0}{\tau b}y = \frac{1}{\tau}EqB \tag{9.8}$$

where,

$$B = \frac{1}{3\pi\mu d_p} \tag{9.9}$$

The initial conditions of above differential equations are written as

$$t = 0,\quad x = x_0,\quad y = b,\quad \omega_x = \frac{dx}{dt} = 0,\quad \omega_y = \frac{dy}{dt} = -u_0 \tag{9.10}$$

The particle trajectory of the particle is obtained by solving the equations (9.7) and (9.8), which is written as

$$x = \frac{1+\alpha}{2\alpha}x_0\exp\left[-\frac{(1-\alpha)t}{2\tau}\right] - \frac{1-\alpha}{2\alpha}x_0\exp\left[-\frac{(1+\alpha)t}{2\tau}\right] \tag{9.11}$$

$$y = \left[\frac{1+\beta}{2\beta}\left(1+\frac{EqB}{u_0}\right)b - \frac{\tau u_0}{\beta}\right]\exp\left[-\frac{(1-\beta)t}{2\tau}\right] - EqB\frac{b}{u_0} \tag{9.12}$$

where,

$$\alpha = \sqrt{1 + 4\tau u_0/b},\quad \beta = \sqrt{1 - 4\tau u_0/b} \tag{9.13}$$

Because the relaxation time τ is very small, the equation (9.13) can be written as

$$\alpha \approx 1 + 2\tau u_0/b,\quad \beta = 1 - 2\tau u_0/b \tag{9.14}$$

Because of $(1+\alpha)\exp\left[-\frac{(1-\alpha)t}{2\tau}\right] \gg (1-\alpha)\exp\left[-\frac{(1+\alpha)t}{2\tau}\right]$, then equation (9.11) can be simplified as

$$x = \frac{1+\alpha}{2\alpha}x_0\exp\left[-\frac{(1-\alpha)t}{2\tau}\right] = x_0\exp\left(-\frac{u_0}{b}t\right) \tag{9.15}$$

Since the relaxation time τ is very small, $\frac{\tau u_0}{\beta} \to 0$, equation (9.12) can be simplified as

$$y = \frac{1+\beta}{2\beta}b\left(1+\frac{EqB}{u_0}\right)\exp\left[-\frac{(1-\beta)t}{2\tau}\right] - EqB\frac{b}{u_0} \tag{9.16}$$

If a particle with d_p is collected at the end of the plate $x = l$, as shown as the dotted line, then in the range between the dotted line (the upper limit trajectory of the particle) and $x \geqslant 0$, $y \geqslant 0$, all the particles with d_p are collected, and the separation width is x_0. A half of the flow section width is a as shown in Fig. 9.2. Therefore, the collection efficiency in laminar flow is given by

$$\eta_l = x_0/a \tag{9.17}$$

Now, the problem is to find x_0. Suppose that t_0 is the time of the particle moving from $y = b$ to $y = 0$ along the dotted line. Let $x = l$, $t = t_0$. From equation (9.15), we obtain

$$x_0 = l\exp\left(-\frac{u_0 t_0}{b}\right) \tag{9.18}$$

In equation (9.16), let $y = 0$, the time of the particle moving from $y = b$ to $y = 0$ is expressed as

$$t_0 = \frac{2\tau}{1-\beta}\ln\left[\frac{1+\beta}{2\beta}\left(1 + \frac{u_0}{BqE}\right)\right] \tag{9.19}$$

According to equation (9.14), equation (9.19) is simplified as

$$t_0 = \frac{b}{u_0}\ln\left(1 + \frac{u_0}{BqE}\right) \tag{9.20}$$

Substituting equation (9.20) into equation (9.18), we have

$$x_0 = l\left(1 + \frac{u_0}{BqE}\right)^{-1} \tag{9.21}$$

Substituting equation (9.21) into equation (9.17), the collection efficiency of the windward side of the plate under the laminar flow is given by

$$\eta = \frac{l}{a}\left(1 + \frac{u_0}{BqE}\right)^{-1} \tag{9.22}$$

Because of $BqE = qE/3\pi\mu d_p = \omega$, and $u_0/\omega \gg 1$, equation (9.22) becomes

$$\eta = \frac{l\omega}{au_0} \tag{9.23}$$

However, in the transverse plate ESP, the flow field is the turbulent flow. According to the relation of the collection efficiency in the laminar flow and the turbulent flow (Chapter 5), the collection efficiency of the windward side of one transverse plate in the turbulent flow is given by

$$\eta_t = 1 - \exp\left(-\frac{l\omega}{au_0}\right) \tag{9.24}$$

The leeward side of the plate also has the collection effect. But it is difficult to derive the collection efficiency theoretically because in the back of the plate, as shown in Fig. 9.2, the gas velocity is much lower than the front of the plate. The collection efficiency of leeside of the plate should be higher than that of the windward side of the plate. For a conservative calculation, the collection efficiency of the two sides of a single plate is given by

$$\eta_1 = 1 - (1 - \eta_t)^2 \tag{9.25}$$

If there are k plates in series in a transverse plate ESP, the overall collection efficiency is given by

$$\eta_0 = 1 - (1 - \eta_t)^{2k} = 1 - \left[\exp\left(-\frac{l\omega}{au_0}\right)\right]^{2k} \tag{9.26}$$

It can be seen that the higher collection efficiency could be achieved as the plate width $2l$ increasing and the flow section width $2a$ decreasing. However, at the same gas flow rate, the average gas velocity in the flow section u_0 will be increased which will lead to the reduction of the collection efficiency. Therefore, the optimization of the geometric sizes of the electrodes in the transverse plate ESP is required.

Example 9.1 A transverse plate ESP with 20 transverse plates in series. A half of a transverse plate width is 0.2m, a half of the flow section width is 0.15m, the average gas velocity in the flow section is 2m/s, and the migration velocities of 1μm and 5μm particles are 0.04m/s and 0.1m/s respectively. Estimate the collection efficiencies of this transverse plate ESP for 1μm and 5μm particles.

Solution

According to equation (9.26), the collection efficiency for 1μm particle is

$$\eta_0 = 1 - \left[\exp\left(-\frac{l\omega}{au_0}\right)\right]^{2k} = 65.6\%$$

The collection efficiency for 5μm particle is

$$\eta_0 = 1 - \left[\exp\left(-\frac{l\omega}{au_0}\right)\right]^{2k} = 93.1\%$$

9.1.2 Electrostatic Enhancement Cyclone

Electrostatic enhancement cyclone is a hybrid separator of cylindrical ESP and cyclone. The schematic configuration of a electrostatic enhancement cyclone is shown in Fig. 9.3.

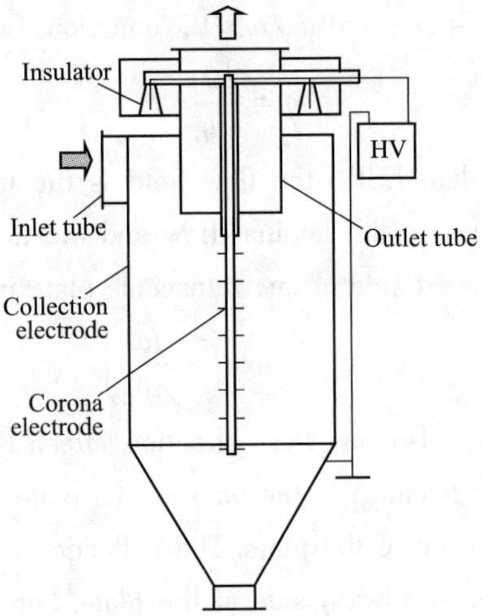

Fig. 9.3 Electrostatic enhancement cyclone

A discharging electrode is fixed in the axes of a tangential inlet reverse-flow cyclone. The cyclone body takes as a function of a tube collection electrode. In this case, both the centrifugal force and electric force well act on the charged particles simultaneously. Thus, the collection efficiency could be further improved[9].

The collection efficiency of an electrostatic enhancement cyclone can be developed based on the cut diameter model in Chapter 5. From equations (2.34) and (5.78), where the Cunningham slip correction factor is not considered, the settling velocity of the particle with the cut diameter d_c toward the wall of the cyclone is given by

$$w_p = \frac{\rho_p d_c^2}{18\mu}\frac{u^2}{r_c} + \frac{qE}{3\pi\mu d_c} \tag{9.27}$$

Assume the charge on a particle is saturated, which is given by

$$q = 3\pi\varepsilon_0 E d_p^2 \left(\frac{\varepsilon}{\varepsilon + 2}\right) \tag{9.28}$$

Substituting equation (9.28) into equation (9.27), we have

$$w_p = \frac{\rho_p d_c^2}{18\mu} \frac{u^2}{r_c} + \left(\frac{\varepsilon}{\varepsilon + 2}\right) \frac{\varepsilon_0 d_c E^2}{\mu} \tag{9.29}$$

Where r_c ——the radius of the cylindrical separation surface;

E ——the field strength at the cylindrical separation surface.

Because at the cylindrical separation surface a particle settling velocity outward is equal to the radical gas velocity inward, we have

$$w_p = \frac{\rho_p d_c^2}{18\mu} \frac{u^2}{r_c} + \left(\frac{\varepsilon}{\varepsilon + 2}\right) \frac{\varepsilon_0 d_c E^2}{\mu} - \frac{Q}{2\pi r_c L} = 0 \tag{9.30}$$

Where L——the natural length, given by equation (5.63).

Thus, the cut diameter can be found by

$$d_c = \frac{9\mu r_c}{\rho_p u^2} \left[\sqrt{\frac{\varepsilon_0 \varepsilon E^2}{(\varepsilon + 2)\mu} + \frac{\rho_p Q u^2}{9\pi\mu L r_c^2}} - \frac{\varepsilon_0 \varepsilon E^2}{(\varepsilon + 2)\mu} \right] \tag{9.31}$$

Where, if the radius of the cylindrical separation surface r_c is assumed to be the radius of the outlet tube r_1, then, according to equation (5.87), the collection efficiency of an electrostatic enhancement cyclone can be predicted by

$$\eta = 1 - \exp[-0.693(d_p/d_c)^2] \tag{9.32}$$

It is a good idea to combine ESP and cyclone to achieve the higher collection efficiency. However, in ESPs, lower velocity is required to get the higher efficiency. But in cyclones, the ranges of about 15m/s to 25m/s are needed in order to achieve high efficiency. These velocities are too high for the operation of ESPs. This contradiction has obstructed the application of the electrostatic enhancement cyclone[10]. If some measures were found to prevent the collection wall from the dust re-entrainment, the electrostatic enhancement cyclone would become one of the practical separators.

9.2 Electrostatic Enhancement Fabric Filters

Electrostatic enhancement fabric filter is a kind of a hybrid electrostatic filtration separator[11]. Collection performances of fabric filters can be enhanced electrically by means of the pre-charging electric field applied on the dust-laden gas to produce charged particles, and the external electric field applied between the fabric filtration median to increase the electric collection effect, or both[12].

An external electric field is seldom employed because it leads to the collection system being more complex and increasing the bag burning risk due to the back corona on the surface of the filter. Therefore, the particle pre-charging fabric filters are more widely used in practice. There are two particle pre-charging techniques including unipolar dust pre-charging and bipolar dust pre-charging.

9.2.1 Unipolar Dust Pre-chargers for Fabric Filters

It has been proved that the filtration efficiency of fiber is obviously improved when the particles are charged[13]. The typical structure of a particle pre-charging fabric filter is shown in Fig. 9.4.

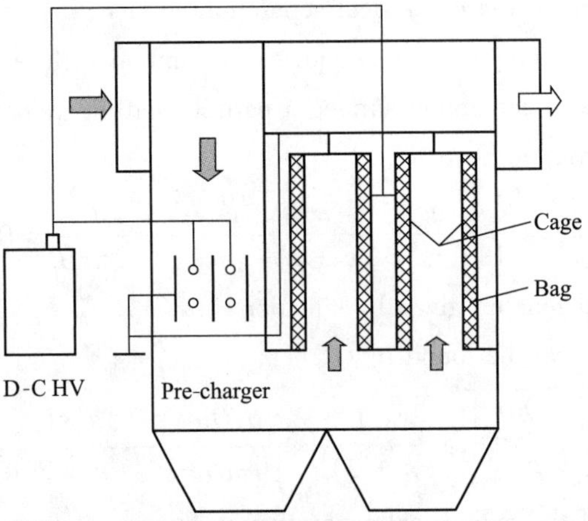

Fig. 9.4 Typical structure of a unipolar dust pre-charging fabric filter

The unipolar particle pre-charger is as the same as the single-stage ESP. Then, a simplest way of unipolar dust pre-charging for increasing the collection efficiency is to install an ESP in the front of the bag house, and usually negatively charged particles, such as the COHPAC developed by EPRI of the United State[14,15], as shown in Fig 9.5.

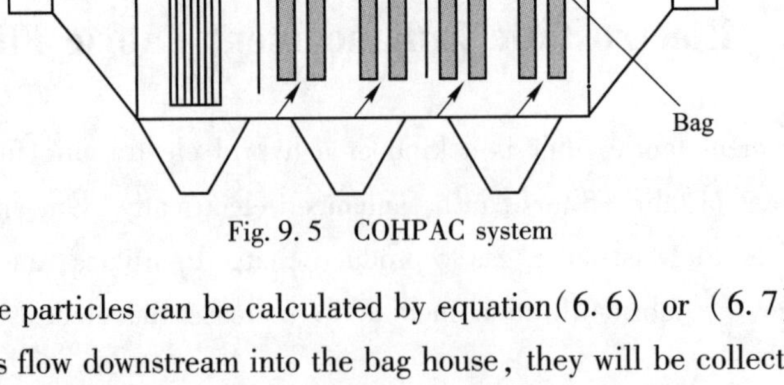

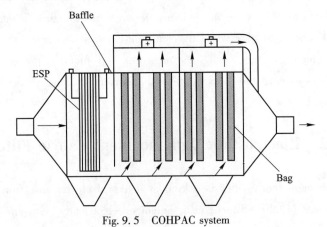

Fig. 9.5 COHPAC system

The charges on the particles can be calculated by equation (6.6) or (6.7). When the charged particles with the gas flow downstream into the bag house, they will be collected by the fabric bag.

If the fibers of the bags are neutral, the fractional collection efficiency of a single fiber has been given by Pich[16] theoretically as

$$\eta_E = \eta_R + \frac{K_I}{G^2} \quad (x \leqslant 1) \tag{9.33}$$

$$\eta_E = 2\sqrt{K_I/H} \quad (x > 1) \tag{9.34}$$

Where η_R ——the interception collection efficiency of a fiber, which is given by equation (7.9);

G——the interception parameter, which is given by equation (7.6).

Other parameters in equations (9.33) and (9.34) are given by

$$x = \frac{K_I}{G^2 \eta_R} \tag{9.35}$$

$$K_I = \frac{\varepsilon_p - 1}{\varepsilon_p + 2}\left(\frac{4C_c q^2 L}{3\pi\mu d_p v}\right) \tag{9.36}$$

Where L ——the thickness of the fabric material;

d_p ——the particle diameter;

v ——the gas velocity;

C_c ——the Cunningham slip correction factor given by equation (1.3).

Since the particles which are larger than 1μm have been removed by the ESP, the inertial impaction collection efficiency of a fiber can be neglected. The combined collection efficiency of a fiber is given by

$$\eta = 1 - (1 - \eta_R)(1 - \eta_D)(1 - \eta_E) \tag{9.37}$$

The overall collection efficiency of a fibrous filter bed can be calculated by equation (7.23).

Example 9.2 The operation data of a unipolar dust pre-charging bag house are given in Table 9.1. Compute the collection efficiency of the felt filter for 1.0μm particle.

Table 9.1 Operation data of an unipolar dust pre-charging bag house

$E_q/V \cdot m^{-1}$	$\varepsilon_0/(CV \cdot m)^{-1}$	ε	ρ /kg · m^{-3}	v /m · min^{-1}	d_f/μm	L/mm	β	μ/Pa · s	$D/m^2 \cdot s^{-1}$
500	8.85×10^{-12}	6	1	1.5	30	3	0.3	1.85×10^{-5}	2.7×10^{-11}

Solution

(1) Interception. The Reynolds number flow around the cylindrical fiber is

$$\text{Re} = \frac{\rho d_f v}{\mu} = 0.05$$

Lamb constant is

$$L_a = 2 - \ln \text{Re} = 5$$

The interception parameter is

$$G = d_p/d_f = 0.033$$

From equation (7.9), the interception efficiency is

$$\eta_R = \frac{1}{L_a}\left[(1 + G)\ln(1 + G) - \frac{G(2 + G)}{2(1 + G)}\right] = 8.9 \times 10^{-4}$$

(2) Diffusion. The Peclet number is

$$\text{Pe} = \frac{d_p v}{D} = 9.3 \times 10^2$$

The diffusion collection efficiency is calculated by equation (7.13)

$$\eta_D = C\frac{1}{(La)^{1/3}}Pe^{-2/3} = 0.018$$

where, $C=2.92$ (Natason[12]).

(3) Electrostatic effect. From equation (6.7), the saturation charge is

$$q = \pi\varepsilon_0 E_q d_p^2\left[\frac{\varepsilon - 1}{\varepsilon + 2}\frac{2}{1 + 2\lambda/d_p} + (1 + 2\lambda/d_p)^2\right] = 3.15 \times 10^{-17}$$

From equation (9.36), we have

$$K_I = \frac{\varepsilon_p - 1}{\varepsilon_p + 2}\frac{4C_c q^2 L}{3\pi\mu d_p v} = 1.7 \times 10^{-24}$$

From equation (9.35), because of

$$x = \frac{K_I}{G^2\eta_R} \ll 1$$

Then, from equation (9.33), the collection efficiency of the electrostatic effect is

$$\eta_E = \eta_R + \frac{K_I}{G^2} \approx \eta_R = 8.9 \times 10^{-4}$$

(4) Combined collection efficiency. From equation (9.37), The combination efficiency is

$$\eta = 1 - (1 - \eta_R)(1 - \eta_D)(1 - \eta_E) = 0.02$$

(5) Collection efficiency of the bag house. From equation (7.23), the collection efficiency of the bag house for 1.0μm is

$$\eta_0 = 1 - \exp\left[-\frac{4\beta\eta L}{\pi(1 - \beta)d_f}\right] = 66.4\%$$

Comparing this result with the example 7.1, we find that the if the particles are unipolar pre-charged, the collection efficiency of the bag house for 1.0μm particle is increased from 62.3% to 66.4%.

However, there are two problems of the unipolar dust pre-charging bag filter including that (1) it is difficult to clean the particles from the bag filter because of the stronger electrostatic adhesion of the unipolar charged particles on the surface of the bag filter[17], and (2) if the particles on the surface of the fabric material cannot be removed timely, the back corona could possibly occur to cause the bag to be burned due to the accumulation of the charges on the fabrics[18], as shown in Fig. 9.6.

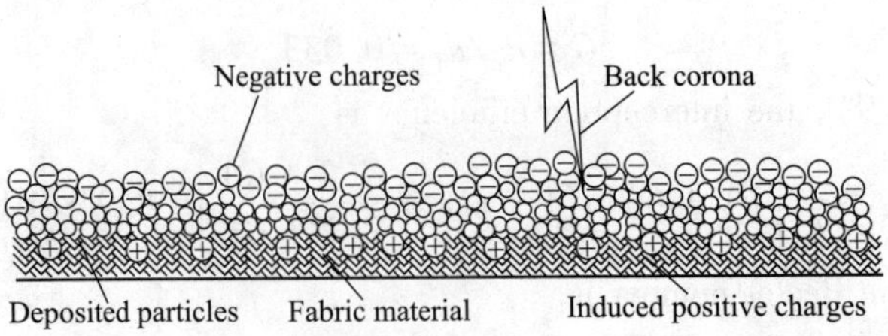

Fig. 9.6　Unipolar charges of particles on the surface of the fabric material

9. 2. 2 Bipolar Dust Pre-chargers for Fabric Filters

If the pre-charging particles are bipolar, the electrostatic neutralization of the charged particles on the surface of the filter occurs. This neutralization effect leads to less charges on the surface of the fabric material. Then, the electrostatic adhesion force of the charged particles on the fabric median is reduced. Furthermore, the bipolar charged particles may abstract each other to form a fluffy agglomerated particle layer. Thus, the dust can be easily removed from the bag filter, as shown in Fig. 9. 7. Obviously, the potential risk of burning bags can be avoided.

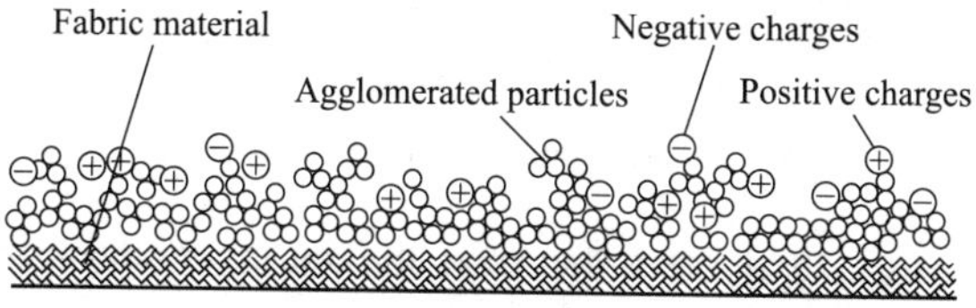

Fig. 9. 7 Bipolar charges of particles on the surface of the fabric material

9. 2. 2. 1 Bipolar Pre-charger with Two High Voltage Supplies

If a pre-charger can produce bipolar charges, two different polarities of high voltage supplies should to used[19], as shown in Fig. 9. 8. When the corona wires are connected to the negative DC voltage supply output, the wires produce the negative ions which impact on the particles to form the negative charged particles. On the other hand, the positive charged particles can be formed by the positive DC voltage supply. Thus, the bipolar charged particles can be obtained in Fig. 9. 8.

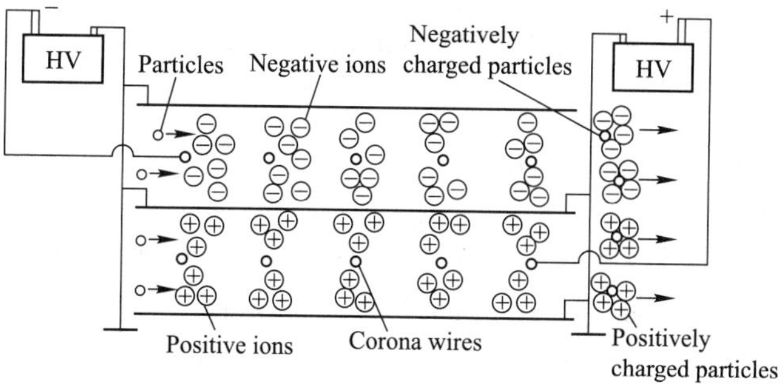

Fig. 9. 8 Bipolar pre-charger with two high voltage supplies

9. 2. 2. 2 Bipolar Pre-charger with One High Voltage Supply

To generate bipolar charges with two high voltage supplies makes the electric system being too complicated. According to the physical principle, in a electric field between two electrodes, the smaller diameter electrode will start corona discharging firstly no matter what output polarity of the high voltage supply is connected to the larger diameter electrode or smaller diameter electrode. Therefore, a new bipolar per-charger, as shown in Fig. 9. 9, can be developed[20].

In Fig. 9. 9, the diameter of the tube electrodes (ranging from 15mm to 40mm) is much larger than that of the corona wires. The negative corona wires are located between the positive tubes,

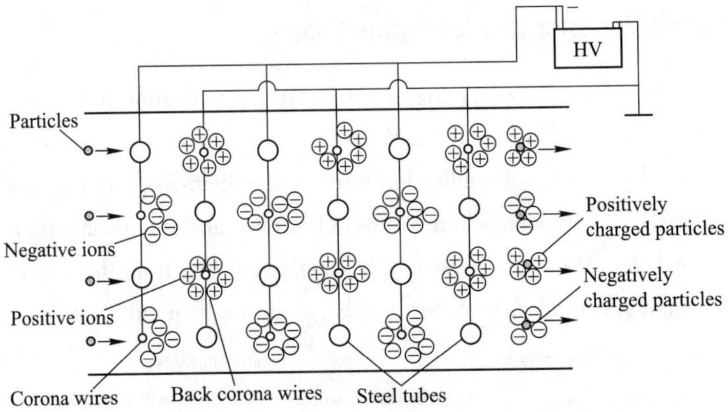

Fig. 9. 9　Bipolar pre-charger with one high voltage supply

and the positive corona wires are located between the negative tubes. In the field between the positive tubes, the particles are negatively charged. At same time, in the field between the negative tubes, the particles are positively charged by the positive corona wires. Thus, bipolar charged particles can be produced by one high voltage supply.

Both bipolar pre-charger with one or with two high voltage supplies can be used to enhance the collection efficiency of a fabric filter. However, recent researches have proved that the efficiency enhancement, the *drag reduction*, *and charge accumulation restraining effects* of the bipolar particle pre-charging is much better than that of the unipolar particle pre-charging in electrostatic enhancement fabric filters[21].

9.3　Electrostatic Scrubbers

9.3.1　Collection Efficiency Prediction

In electrostatic scrubber, the key device is the charged spray generator, or droplet pre-charger[22, 23]. The corona discharging method is the most commonly used for water drops charging in application. Fig. 9. 10 is a kind of electrostatic spray nozzle, and Fig. 9. 11 is a other simple way of the charged spray generation.

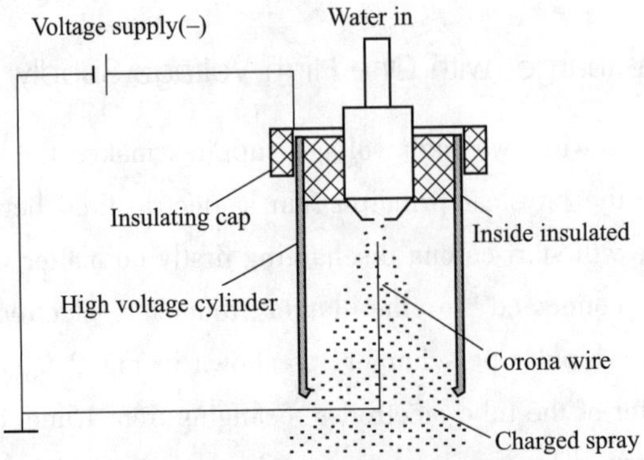

Fig. 9. 10　Electrostatic spray nozzle

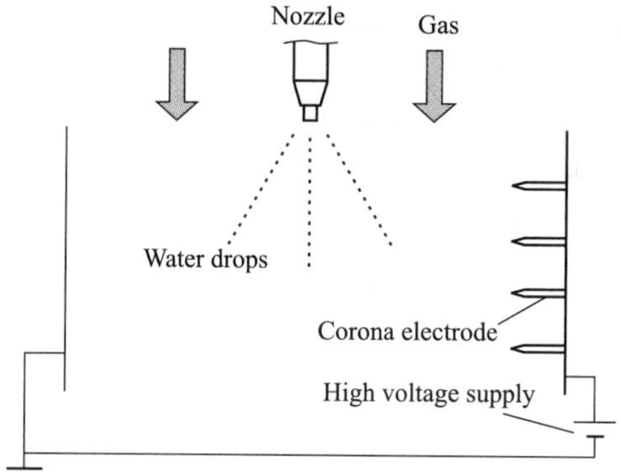

Fig. 9.11 Generation of charged water drops in electric field

We assume that the charges on a droplet can get to the saturation charge very quickly in pre-charging electric field, and as soon as the particles contact the water drops, they will be captured by water drops. The charges on the droplets can be calculated by equation (6.6).

If the particles are neutral, the charge on a drop is Q_w. The collection efficiency caused by the induced electric force has been given by Pich[16] as

$$\eta_{E_I} = \left(\frac{15\pi}{8} \frac{\varepsilon_p - 1}{\varepsilon_p + 2} \frac{2C_c d_p^2 Q_w^2}{3\pi\mu d_w v_0 \varepsilon_0} \right)^{0.4} \tag{9.38}$$

It is can be seen from equation (9.38) that the collection efficiency caused by the induced electric force is very small. If the changes on a particle is q, the charge on a drop is Q_w, and their polarities are different. The collection efficiency caused by the electrostatic attraction force or Coulomb force is given by

$$\eta_{E_c} = \frac{4C_c Q_w q}{3\pi^2 \varepsilon_0 \mu d_p d_w^2 v_0} \tag{9.39}$$

The enhancement effect of the Coulomb force is greater than that of the induced electric force. When the particles and water drops in the electric field E_0, the collection efficiency caused by the electric field force is given by

$$\eta_{E_0} = \frac{K_E}{1 + K_E} \left(1 + 2 \frac{\varepsilon_w - 1}{\varepsilon_w + 2} \right), \quad K_E = \frac{C_c q E_0}{3\pi^2 \mu d_p v_0} \tag{9.40}$$

Where ε_w——the relative permittivity of water.

The electric field force has a stronger enhancement effect on the electrostatic scrubbers. The combined collection efficiency of a drop is written as

$$\eta = 1 - (1 - \eta_R)(1 - \eta_D)(1 - \eta_E) \tag{9.41}$$

where, η_E presents η_{E_i}, η_{E_c}, or η_{E_0}, which depends on the electric forces acting on the particles.

Usually the collection efficiency η_{E_i} caused by induced electric force is neglected.

In a electrically charged spray tower, the overall collection efficiency can be predicted by equation (8.27).

9. 3. 2 Electrically Charged Spray Tower

Fig. 9. 12 is a electrically charged spray tower. The charged water droplets are formed in the electric fields by wire-plate electrodes. When the dirty gas gets into the tower to mix with the charged droplets, the particles in the gas will be collected by charged droplets[24]. The dirty water drops then fall to the waste water tank at the bottom of the tower. The clean gas is discharged outside from the outlet tube.

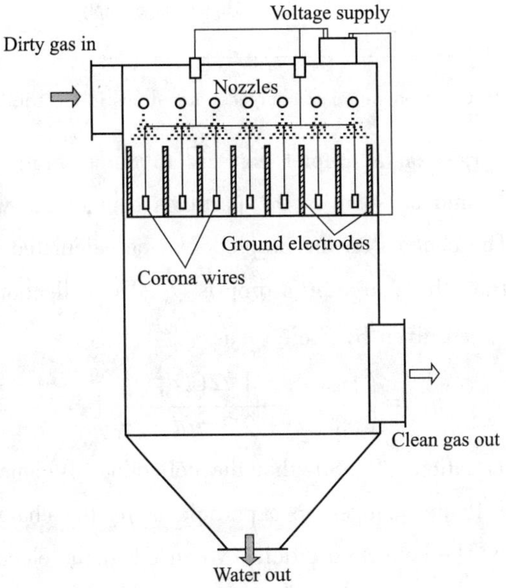

Fig. 9. 12 Electrically charged spray tower

It is noticed that the gas flows in from the top of the tower. The distinguish advantage is that the insulators are kept in dry surroundings because they are located in the hot gas area. Thus, it is not necessary to assemble a heating system for insulators.

Although the collection efficiencies of charged droplets for the particulate pollutant have been given theoretically in a electrically charged spray tower, it is still very difficult to predicted the enhancement effect of the charged droplets precisely because there are many the effect factors which could not be determined quantitatively, such as the effects of the humidity, gas temperature, compositions, dust properties, and the size distributions of particles and droplets.

9. 3. 3 Wet Electrostatic Precipitators

Wet electrostatic precipitators (WESPs) are widely employed for the collection of fine particulates, oil, smoke and acid mist[25-27]. There are two types of WESPs, horizontal-flow and vertical-flow precipitators. The collection efficiency of a wet ESP for fine particles is much higher than that of a dry ESP due to less dust re-entrainment and better charging property of wetted particles[28].

Fig. 9. 13 shows a cross sectional overview of a horizontal-flow WESP. The horizontal-flow

WESPs are generally employed for large gas flow rate treatment. The electrode arrangement of WESPs is as the same as the wire-plate ESPs for the particulate separation in the dry gas.

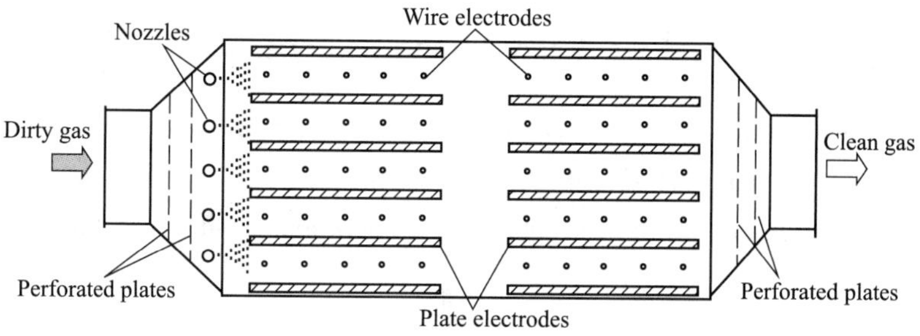

Fig. 9.13 Horizontal-flow WESP

Because of the high moisture gas inside the WESP, more attention has to be paid on the high voltage insulation. The high voltage insulators must kept dry and clean by adding a heating system and a purge-air system.

Fig. 9.14 is a schematic vertical-flow WESP. The vertical-flow precipitators are usually used for lower gas flow rate. If the vertical-flow precipitators are used for controlling the large gas flow rates, the parallel connections are needed.

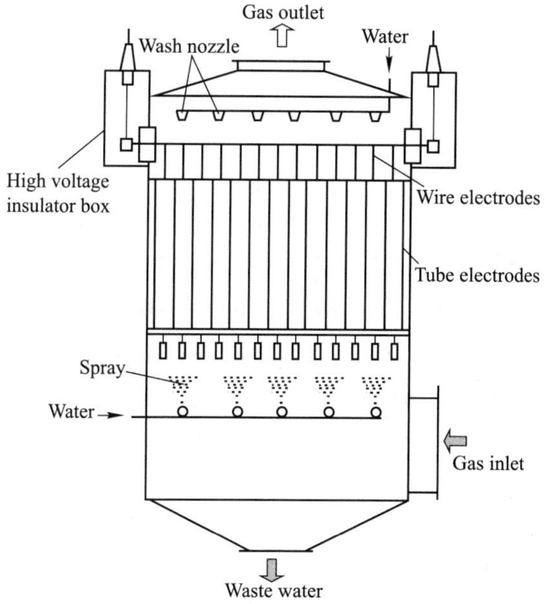

Fig. 9.14 Vertical-flow WESP

The hexagonal tubes are usually used as the collection electrodes in the applications[29]. Typically the flue gases enter at the bottom and rise through the precipitator[30]. There are generally two sets of spray nozzles. The first set at the bottom continually cools and saturate the flue gases. Another spray nozzle set at the top washes down the electrodes periodically to clean the scales in the WESP.

9.4　Aerodynamic Free Rotary Thread Scrubber

9.4.1　Working Principal of Free Rotary Thread Scrubber

The free rotary thread scrubber which was recently developed by Xiang[31] can be used to remove the particles in dry gas and the liquid droplets by means of many high speed turning fabric threads. A schematic diagram of the free rotary thread scrubber is shown in Fig. 9.15.

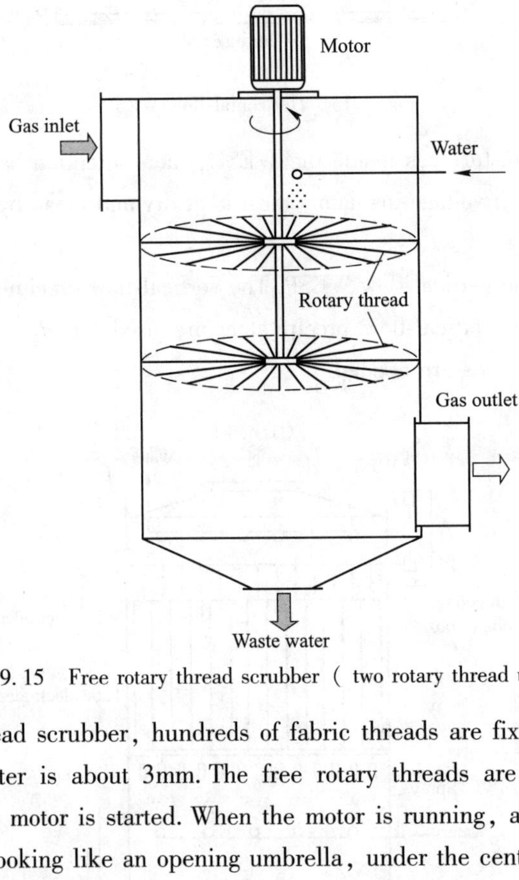

Fig. 9.15　Free rotary thread scrubber (two rotary thread units)

In the free rotary thread scrubber, hundreds of fabric threads are fixed on a motor connecting shaft. The thread diameter is about 3mm. The free rotary threads are pendent naturally under gravity action before the motor is started. When the motor is running, all the flexible fiber rotary threads are stretched, looking like an opening umbrella, under the centrifugal force action.

In some cases, the collection efficiency may be not satisfied the requirements of the pollutant emission concentrations if only one layer of rotary threads is fixed inside the free rotary thread scrubber because the number of the threads in one layer is limited. In order to arrange more rotary threads to enhance the efficiency, two or more layers of fabric threads are installed in a free rotary thread scrubber. In Fig. 9.15, two rotary thread layers, which is called as two rotary thread units, are used. It is obvious that the pressure drop of the free rotary thread scrubber is very low since the gas can pass easily through the gaps of the free rotary threads.

When the particle-containing gas flows into the scrubber, the particles will pass through the section of the rotary thread units and collide with the rotary threads. Then the particles in the gas will be captured by the wet threads. The dust-carrying liquid on the wet threads under the centrifugal force are thrown along the fiber rotary threads to the inner wall of the cylindrical tower.

Furthermore, if some of the particles are not collected by the wet threads, they could be possibly separated by centrifugal force due to the vortex flow which is caused by the thread stirring.

When the free rotary thread scrubber is used to treat the dry pollutant gas, the water must be supplied to keep the threads wetting to make use of the water absorbing property of the fabric material to capture the particulate. Because the fabric threads are flexible, they will be shaking when the thread unit is turning. Therefore, the threads will not be easily scaled. To remove the particles from the dry gas, there is an advantage of that the mist emission concentration of the free rotary thread scrubber is very low because the less liquid-gas ratio is needed.

If the free rotary thread scrubber is used to separate the liquid drops in the wet flue gas, it is often not necessary to feed water for wetting the threads.

The main advantages of the free rotary thread scrubber including that: (1) the inertial impaction separation effect is very strong because the relative speed between the rotary threads and the particles may be as high as tens of meters per second, (2) the high-speed rotating threads can also create spinning gas stream to further increase the particle separation efficiency under the aerodynamic effect, (3) the particle removal ability of the free rotary thread scrubber is very adjustable by means of increasing the number of threads or the rotating speed of the thread driven motor, and (4) the free rotary thread scrubber is a simple structure, low cost, less power consumption, and high efficiency device for separating the particles and mists from gases.

9.4.2 Collection Efficiency Prediction

It can be seen that the collection mechanisms of the free rotary thread scrubber are the combination of the fabric filtration mechanism and the centrifugal separation mechanism. Therefore, the mechanisms of the fabric filtration, just like Fig. 7.1, are consist of interception, inertial impaction, and diffusion. As for the centrifugal separation effect due to the vortex gas flow, the mechanism can be described as the same as the gas flow in arch duct as described in section 5.2.

9.4.2.1 Collection Efficiency of the Free Rotary Threads

The free rotary threads have the filtration collection effects in both vertical and radical directions. Because the collection effect in the vertical direction is negligibly small, we will just discuss the collection efficiency in radical direction.

Interception

The gas velocity around the thread is quite high (hundreds of times greater than that of fabric filter), and the diameter of the thread is large, usually in the range from 1mm to 4mm in diameter. It can be predicted that the Reynolds number is greater than 100. Then the flow field belongs to the potential flow because of $Re \gg 1$.

In potential flow, the interception efficiency can be described by equation (7.5). Because the interception parameter G in equation (7.6) is very small, the interception efficiency in an uniform gas flow can be simplified as

$$\eta_R = 2G, \qquad G = d_p / d_f \tag{9.42}$$

Where d_p, d_f——the particle diameter and the thread diameter.

However, the radical velocity u along a thread, as shown in Fig. 9. 16 is not uniform. The mass of particles captured by a element length of dr on the thread due to the interception given by

$$dm = c_0 \eta_R d_f u dr = c_0 \eta_R d_f \omega r dr \tag{9.43}$$

Where c_0——the particle concentration in gas;

ω——the angular velocity in rad/s.

Integrating equation (9.43), the collected mass of particles on a single thread is expressed as

$$m = \int_0^{r_0} c_0 \eta_R d_f \omega r dr = \frac{1}{2} c_0 \eta_R d_f \omega r_0^2 \tag{9.44}$$

Where r_0——the length of a thread.

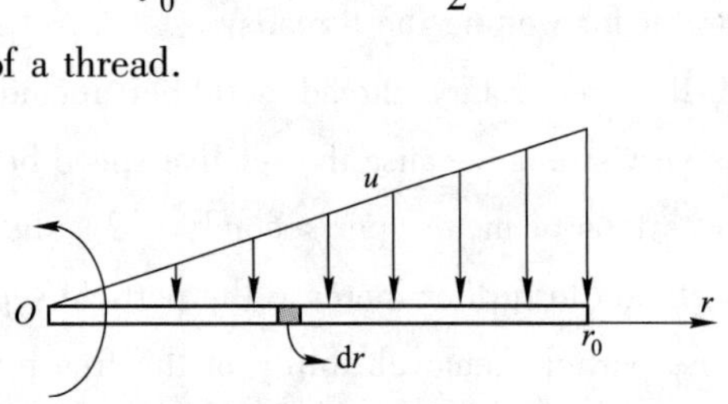

Fig. 9. 16 Efficiency development model of a single free rotary thread

If there are N threads, the total collected mass is given by

$$m_N = \frac{1}{2} N c_0 \eta_R d_f \omega r_0^2 \tag{9.45}$$

The total mass of the particles flows to the threads is expressed as

$$M = Qc_0 = \pi r_0^2 v c_0 \tag{9.46}$$

Where v——the average vertical gas velocity in the free rotary thread scrubber.

Then, the interception efficiency of N threads is given by

$$\eta_{RN} = \frac{m_N}{M} = \frac{\frac{1}{2} N c_0 \eta_R d_f \omega r_0^2}{\pi r_0^2 v c_0} = \frac{\omega}{2\pi v} N d_f \eta_R \tag{9.47}$$

Inertial Impaction

For the potential flow, the inertial impaction collection efficiency of a single fiber in the uniform distributed velocity is given by equation (7.11), which is rewritten as

$$\eta_I = \frac{\mathrm{St}^3}{\mathrm{St}^3 + 0.77\mathrm{St}^2 + 0.22} \tag{9.48}$$

We can use the same method to develop the inertial impaction collection efficiency of the free rotary threads when the velocity is $u = \omega r$, as shown in Fig. 9. 16. The collected mass of particles on a single thread due to the inertial impaction effect expressed as

$$m = \int_0^{r_0} c_0 \eta_I d_f \omega r dr \tag{9.49}$$

However, because the Stokes number in equation (9.48) is the function of u, it is given by

$$\mathrm{St}=\frac{\rho_{\mathrm{p}}d_{\mathrm{p}}^{2}u}{18\mu d_{\mathrm{f}}} \tag{9.50}$$

Therefore, we have to use the numerical integration method to solve the equation (9.49). It will be too complex to calculate the inertial impaction efficiency when the gas velocity is not uniform toward the threads. For an approximate prediction, the average velocity is introduced by

$$\bar{u}=\frac{1}{r_0}\int_0^{r_0}\omega r\mathrm{d}r=\frac{1}{2}\omega r_0 \tag{9.51}$$

Then, the Stokes number is given by

$$\mathrm{St}=\frac{\rho_{\mathrm{p}}d_{\mathrm{p}}^{2}\bar{u}}{18\mu d_{\mathrm{f}}} \tag{9.52}$$

Thus, the inertial impaction efficiency of N threads can be derived by the same method as the interception efficiency. The result is given by

$$\eta_{\mathrm{IN}}=\frac{\omega}{2\pi v}Nd_{\mathrm{f}}\eta_{\mathrm{I}} \tag{9.53}$$

Diffusion

When a particle diameter is less than 0.5μm, the diffusion effect may be not negligible. For the potential flow, the diffusion efficiency had been given by Natason[32] as

$$\eta_{\mathrm{D}}=3.19/\sqrt{\mathrm{Pe}}=3.19\Big/\left(\frac{d_{\mathrm{p}}u}{D}\right)^{1/2} \tag{9.54}$$

Then the collected mass of particles on the free rotary threads at the number of N due to the diffusion is expressed as

$$m_{\mathrm{DN}}=N\int_0^{r_0}c_0\eta_{\mathrm{D}}d_{\mathrm{f}}\omega r\mathrm{d}r \tag{9.55}$$

Substituting equation (9.54) into equation (9.55), we have

$$m_{\mathrm{DN}}=2.13Nc_0(D\omega/d_{\mathrm{p}})^{1/2}d_{\mathrm{f}}r_0^{3/2} \tag{9.56}$$

Then, the diffusion efficiency of N threads is expressed as

$$\eta_{\mathrm{DN}}=\frac{m_{\mathrm{N}}}{M}=\frac{2.13Nc_0(D\omega/d_{\mathrm{p}})^{1/2}d_{\mathrm{f}}r_0^{3/2}}{\pi r_0^{2}vc_0}=\frac{2.13(D\omega/d_{\mathrm{p}}r_0)^{1/2}}{\pi v}Nd_{\mathrm{f}} \tag{9.57}$$

9.4.2.2 Centrifugal Separation Efficiency of the Vortex Flow in Free Rotary Thread Scrubber

In a free rotary thread scrubber, two or more rotary thread units are usually arranged. In this way higher collection efficiency can be obtained because more threads are used. At the same time, a quite perfect vortex gas flow can be formed between thread units to enhance the aerodynamic separation effect.

Suppose that two free rotary thread units are connected in series in the free rotary thread scrubber, as shown in Fig. 9.17.

The distance h between two thread units is often less than or equal to the length r_0 of a

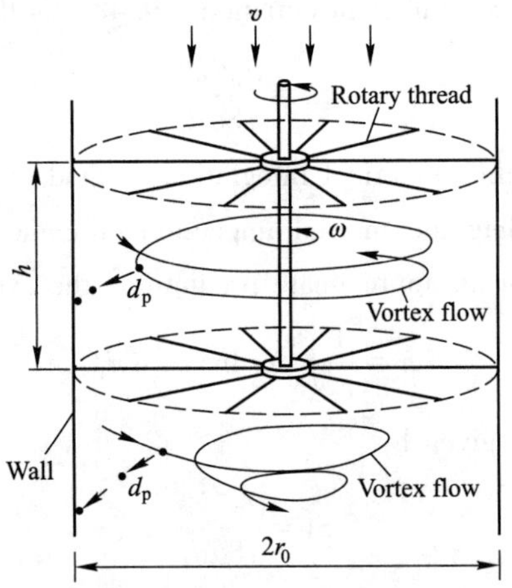

Fig. 9. 17 Particle or droplet separations under the centrifugal force in a free rotary thread scrubber

thread. The fast turning threads will force the block of the gas in the space h turning almost at the same rotating speed of the rotary threads, written as

$$u = \omega r \tag{9.58}$$

The time of the gas moving downward is given by

$$t_0 = h/v \tag{9.59}$$

In this period, the total angle of the vortex flow is given by

$$\theta = \omega t_0 = \omega h/v \tag{9.60}$$

Suppose that a particle of d_p in diameter is located at the radius r_θ. When a particle moves an angle θ with the flow and deposits on the wall of the duct at the radius $r = r_0$, this particle is separated, as shown as the dotted line in Fig. 9. 18.

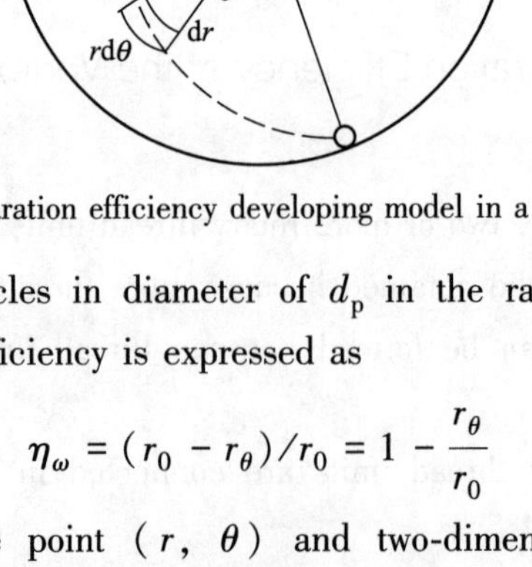

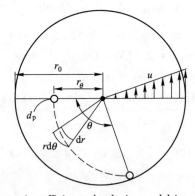

Fig. 9. 18 Centrifugal separation efficiency developing model in a free rotary thread scrubber

It is clear that all the particles in diameter of d_p in the range between r_θ and r_0 will be collected. Thus the collection efficiency is expressed as

$$\eta_\omega = (r_0 - r_\theta)/r_0 = 1 - \frac{r_\theta}{r_0} \tag{9.61}$$

An element of fluid at the point (r, θ) and two-dimensional flow are considered in Fig. 9. 18. The particle moving distances in radial direction and tangential direction in time dt are written as

$$dr = w\mathrm{d}t, \qquad r\mathrm{d}\theta = u\mathrm{d}t \tag{9.62}$$

From equation (9.62), we have

$$\frac{\mathrm{d}r}{\mathrm{d}\theta} = r\frac{w}{u} \tag{9.63}$$

The centrifugal accelerating velocity of the particle is [shown in equation (2.31)] given by

$$w = \tau\frac{u^2}{r} \tag{9.64}$$

where, τ determined by equation (2.13).

Then equation (9.63) becomes

$$\mathrm{d}r = \tau u\mathrm{d}\theta \tag{9.65}$$

Substituting equation (9.58) into equation (9.65), we have

$$\mathrm{d}r = \tau r\omega\mathrm{d}\theta \tag{9.66}$$

Integrating equation (9.66), we have

$$\int_{r_\theta}^{r_0}\frac{\mathrm{d}r}{r} = \int_0^\theta \tau\omega\mathrm{d}\theta \tag{9.67}$$

Thus, we obtain

$$r_\theta = r_0\mathrm{e}^{-\omega\tau\theta} \tag{9.68}$$

Substituting equation (9.68) into equation (9.61), then we obtain the centrifugal separation efficiency of the vortex flow in free rotary thread scrubber, written as

$$\eta_\omega = 1 - \frac{r_\theta}{r_0} = 1 - \mathrm{e}^{-\omega\tau\theta} \tag{9.69}$$

Substituting equation (9.60) into equation (9.69), the centrifugal separation efficiency can also be written as

$$\eta_\omega = 1 - \exp\left(-\frac{\tau h\omega^2}{v}\right) \tag{9.70}$$

It can be seen that if the rotary thread rotating speed ω is increased, the centrifugal separation efficiency will be obviously increased. In this case, the power of the thread driven motor will also be tremendously increased either.

Again, beneath the second free rotary thread unit, there is also the centrifugal separation effect, as shown in Fig. 9.17. Even though the flow field is too complex because the rotating speed of the vortex gas flow is shrinking as the gas going down, it can be estimated that the centrifugal separation efficiency is about a half of the collection efficiency in the range of h between two thread units, given by

$$\hat{\eta}_\omega \approx 1 - \exp\left(-\frac{\tau h\omega^2}{2v}\right) \tag{9.71}$$

Equation (9.71) is also the centrifugal separation efficiency of a scrubber with only one free rotary thread unit. Then, the total centrifugal separation efficiency is given by

$$\eta_{\omega_\mathrm{T}} = 1 - (1 - \eta_\omega)(1 - \hat{\eta}_\omega) \tag{9.72}$$

Example 9.3 Calculate the collection efficiency of a Free rotary thread scrubber as shown in Fig. 9.15 for 5μm particle at standard conditions. The calculating data is given in Table 9.2.

Table 9.2 Operation data of a spray tower

r_0/m	d_f/mm	N	h/m	v/m · s^{-1}	ω/rad · s^{-1}	ρ/kg · m^{-3}	ρ_p/kg · m^{-3}
0.2	3	200	0.2	1	20π	1	2×10^3

Solution

(1) Filtration efficiency of the free rotary threads. According to equation (9.47), the interception efficiency is

$$\eta_{RN} = \frac{\omega}{2\pi v} N d_f \eta_R = 0.10$$

The average the velocity along a thread is

$$\bar{u} = \frac{1}{2}\omega r_0 = 6.28(\text{m/s})$$

The Stokes number is

$$\text{St} = \frac{\rho_p d_p^2 \bar{u}}{18\mu d_f} = 0.314$$

The inertial impaction collection efficiency of a single thread is

$$\eta_I = \frac{\text{St}^3}{\text{St}^3 + 0.77\text{St}^2 + 0.22} = 0.095$$

According to equation (9.53), the inertial impaction efficiency of N threads is

$$\eta_{IN} = \frac{\omega}{2\pi v} N d_f \eta_I = 0.570$$

Since the particle is 5μm, the diffusion effect is negligible. The filtration efficiency for one free rotary thread unit is

$$\eta_1 = 1 - (1 - \eta_{RN})(1 - \eta_{IN}) = 0.613$$

Because there are two free rotary thread units in this scrubber, the collection efficiency is

$$\eta_2 = 1 - (1 - \eta_1)^2 = 0.850$$

(2) Centrifugal separation efficiency of the vortex flow. The relaxation time is

$$\tau = \frac{\rho_p d_p^2}{18\mu} = 1.5 \times 10^{-4}\text{s}$$

According to equation (9.70), the centrifugal separation efficiency in the space of h is

$$\eta_\omega = 1 - \exp\left(-\frac{\tau h \omega^2}{v}\right) = 0.112$$

From equation (9.71), the centrifugal separation efficiency contributed by of the vortex flow beneath the second free rotary thread unit is

$$\hat{\eta}_\omega = 1 - \exp\left(-\frac{\tau h \omega^2}{2v}\right) = 0.057$$

The total centrifugal separation efficiency is

$$\eta_{\omega_T} = 1 - (1 - \eta_\omega)(1 - \hat{\eta}_\omega) = 0.162$$

Then, the collection efficiency of a free rotary thread scrubber with two rotary thread units for 5μm particle is

$$\eta = 1 - (1 - \eta_2)(1 - \eta_{\omega_T}) = 0.874 = 87.4\%$$

Exercises

9.1 How many kinds of hybrid separators is described in this chapter? Try to give another kind of hybrid separator.

9.2 In Fig. 9.2, if the flow rate Q, the electrical migration velocity ω, and a half of plate width l are given, and give the relation of optimization value of a half of the flow section width a and l, which will lead to a maximum value of the collection efficiency of the transverse plate ESP, as shown in Fig. 9.1.

9.3 A particle is 5μm in diameter at the charge of 1×10^{-16}C, and a fiber is 20μm in diameter and 1m in length at the charge of 5×10^{-9}C. Determine the largest dust cleaning force [where the distance between is $r = (d_p + d_f)/2$, and permittivity of free space is $\varepsilon_0 = 8.85\times10^{-12}$C/(V · m)].

9.4 It is known that the particles are charged, and the fabric filter is neutral. The relative permittivity of particle is 6, the filter thickness is 2mm, filter diameter is 30μm, and filter porosity is 0.7. Calculate the collection efficiency of the filter for 1μm particle at the charge of 1×10^{-17}C.

9.5 The data are given as a spherical aerosol particle of 1μm with charge of 1×10^{-17}C, a water droplet of 1mm with charge of 1×10^{-14}C, and filtration velocity of 0.05m/s. Calculate the collection efficiency due to the Coulomb force.

References

[1] Jaworek A, Sobczyk A T, Krupa A, et al. Hybrid electrostatic filtration systems for fly ash particles emission control. A review [J]. Separation and Purification Technology, 2019, 213: 283-302.

[2] Xiang X, Chang Y, Nie Y, et al. Investigation of the performance of bipolar transverse plate ESP in the sintering flue control [J]. Journal of Electrostatics, 2015, 76: 18-23.

[3] Fo O B, Marrajr W D, Kachan A G, et al. Filtration of electrified solid particles [J]. Industrial & Engineering Chemistry Research, 2000, 39 (10): 3884-3895.

[4] Masuda S. *Electrostntic Precipitation*, *Handbook of Electrostntic Process* [M]. New York: Marvel Dekker, 1997.

[5] Zou Y P, Zhou Y G, Zhang G Y. Study on the migration velocity of transverse plate ESP [J]. Chinese Journal of Environmental Engineering, 1990, 8 (1): 26-29.

[6] Wu Z F, Zhang G Q, Colbeck I, et al. A theoretical study on the additional velocity of a charged particle due to the impact of ions with high velocity in the collection space of an electrostatic precipitator [J]. Journal of Aerosol Science, 1990, 21: 707-710.

[7] Yi C W, Dou P, Wu C D, et al. Experimental study of transportation characteristic of charged particle in a laboratory scale high velocity electrostatic precipitator, 2010 International Conference on Mechanic Automation and Control Engineering (MACE), Wuhan, 2010: 2038-2042.

[8] Kundu P. *Fluid Mechanics 5th Ed* [M]. Cambridge: Cambridge University Press, 2013.

[9] Lim K S, Kim H S, Lee K W, et al. Comparative performances of conventional cyclones and a double cyclone with and without an electric field [J]. Journal of Aerosol Science, 2004, 35 (1): 103-116.

[10] Chen C. Enhanced collection efficiency for cyclone by applying an external electric field [J]. Separation Science and Technology, 2001, 36 (3): 499-511.

[11] Jaworek A, Krupa A, Czech T, et al. Modern electrostatic devices and methods for exhaust gas cleaning: A brief review [J]. Journal of Electrostatics, 2007, 65 (3): 133-155.

[12] Wang C. Electrostatic forces in fibrous filters. A review. [J]. Powder Technology, 2001, 118 (1): 166-170.

[13] Ardkapan S R, Johnson M S, Yazdi S, et al. Filtration efficiency of an electrostatic fibrous filter: Studying filtration dependency on ultrafine particle exposure and composition [J]. Journal of Aerosol Science, 2014, 72 (12): 14-20.

[14] Chang R. Compact hybrid particulate collector (COHPAC), Patent US 5158580, 1992.

[15] Chang R. COHPAC compacts emission equipment into smaller, denser unit [J]. Power Engineering, 1996, 100 (7): 22-25.

[16] Borra J P. Review on water electro-sprays and applications of charged drops with focus on the corona-assisted cone-jet mode for High Efficiency Air Filtration by wet electro-scrubbing of aerosols [J]. Journal of Aerosol Science, 2018, 125: 208-236.

[17] Yao Q, Li S, Song Q, et al. Research progress of the control technology of the PM_{10} from combustion sources, 11th International Conference on Electrostatic Precipitation, 20-24 Oct. 2008, Hangzhou, China, 2008, pp. 201-205.

[18] Jaworek A, Czech T, Rajch E, et al. Laboratory studies of back-discharge in fly ash [J]. Journal of Electrostatics, 2006, 64 (5): 326-337.

[19] Ciach T, Sosnowski T R. Removal of soot particles from Diesel exhaust [J]. Journal of Aerosol Science, 1996, 27: S705-S706.

[20] Xiang X D, Li M L, Jia S Y, et al. Charge accumulation restraining effect of bipolar charged particles on fabrics [J]. Chinese Journal of Environmental Engineering, 2018, 38 (8): 2282-2287.

[21] Li X E, Xiang X D, Li M L, et al. Efficiency enhancement and drag reduction effects of bipolar electrostatic bag filter [J]. Chinese Journal of Environmental Engineering, 2019, 39 (1): 141-146.

[22] Natale F D, Carotenuto C, Daddio L, et al. Capture of fine and ultrafine particles in a wet electrostatic scrubber [J]. Journal of environmental chemical engineering, 2015, 3 (1): 349-356.

[23] Carotenuto C, Natale F D, Lancia A, et al. Wet electrostatic scrubbers for the abatement of submicronic particulate. [J]. Chemical Engineering Journal, 2010, 165 (1): 35-45.

[24] Ferhat M F, Ghezzar M R, Smail B, et al. Conception of a novel spray tower plasma-reactor in a spatial post-discharge configuration: Pollutants remote treatment [J]. Journal of Hazardous Materials, 2017, 321: 661-671.

[25] Anderlohr C, Brachert L, Mertens J, et al. Collection and Generation of Sulfuric Acid Aerosols in a Wet Electrostatic Precipitator [J]. Aerosol Science and Technology, 2015, 49 (3): 144-151.

[26] Yang Z, Zheng C, Liu S, et al. A combined wet electrostatic precipitator for efficiently eliminating fine particle penetration [J]. Fuel Processing Technology, 2018, 180: 122-129.

[27] Mertens J, Khakharia P M, Rogiers P, et al. Prevention of mist formation in amine based carbon capture: field testing using a wet electrostatic precipitator (WESP) and a gas-gas heater (GGH) [J]. Energy Procedia, 2017, 114: 987-999.

[28] Yang Z, Zheng C, Chang Q, et al. Fine particle migration and collection in a wet electrostatic precipitator [J]. Journal of the Air & Waste Management Association, 2017, 67 (4): 498-506.

[29] Najafabadi M M, Tabrizi H B, Aramesh A, et al. Effects of geometric parameters and electric indexes on

performance of a vertical wet electrostatic precipitator [J]. Journal of Electrostatics, 2014, 72 (5): 402-411.

[30] Natale F D, Carotenuto C, Daddio L, et al. Capture of fine and ultrafine particles in a wet electrostatic scrubber [J]. Journal of environmental chemical engineering, 2015, 3 (1): 349-356.

[31] Xiang X D, Zhang Y M, et al. Multi-tube parallel free rotary thread demisting device [P], 15/974539, 2018.

[32] Michael J M, Clyde O. *Filtration-Principles and Practices* [M]. Marcel Dekker, Inc. New York and Basel, 1987.